Conservation and Sustainability in Historic Cities

Dennis Rodwell

Blackwell
Publishing

© text and photographs 2007 by Dennis Rodwell

Blackwell Publishing editorial offices:
Blackwell Publishing Ltd, 9600 Garsington Road, Oxford OX4 2DQ, UK
  Tel: +44 (0)1865 776868
Blackwell Publishing Inc., 350 Main Street, Malden, MA 02148-5020, USA
  Tel: +1 781 388 8250
Blackwell Publishing Asia Pty Ltd, 550 Swanston Street, Carlton, Victoria 3053, Australia
  Tel: +61 (0)3 8359 1011

The right of the Author to be identified as the Author of this Work has been asserted in accordance with the Copyright, Designs and Patents Act 1988.

All rights reserved. No part of this publication may be reproduced, stored in a retrieval system, or transmitted, in any form or by any means, electronic, mechanical, photocopying, recording or otherwise, except as permitted by the UK Copyright, Designs and Patents Act 1988, without the prior permission of the publisher.

First published 2007 by Blackwell Publishing Ltd

ISBN 978-1-4051-2656-4

Library of Congress Cataloging-in-Publication Data

Rodwell, Dennis.
  Conservation and sustainability in historic cities / Dennis Rodwell.
  —1st ed.
    p. cm.
  Includes bibliographical references and index.
  ISBN 978-1-4051-2656-4 (pbk. : alk. paper)
  1. Architecture—Conservation and restoration. 2. Sustainable architecture. 3. Historic preservation. 4. Cultural property—Protection. 5. Urban renewal. I. Title
  NA105.R65 2007
  720.28′8—dc22
                                                    2006036885

Set in 10/12.5 pt Sabon
by Newgen Imaging Systems (P) Ltd, Chennai, India
Printed and bound in Singapore
by Fabulous Printers Pte Ltd

The publisher's policy is to use permanent paper from mills that operate a sustainable forestry policy, and which has been manufactured from pulp processed using acid-free and elementary chlorine-free practices. Furthermore, the publisher ensures that the text paper and cover board used have met acceptable environmental accreditation standards.

For further information on Blackwell Publishing, visit our website:
www.blackwellpublishing.com/construction

Contents

Introduction vii

Acknowledgements viii

Definitions: Conservation and Sustainability ix

1. Conservation: Background — 1
2. Urban Planning Context — 23
3. Sustainability: Background — 47
4. Conservation: International Initiatives and Directions — 64
5. Conservation: United Kingdom Position and Directions — 86
6. Sustainable Cities and Urban Initiatives — 111
7. Managing World Heritage Cities: United Kingdom — 133
8. Managing Historic Cities: the Bottom-Up Approach — 161
9. The Coincidence between Conservation and Sustainability — 183
10. The Challenge and the Opportunity — 204

Sources 217

Bibliography 243

List of Figures 251

Index 255

Introduction

Urban conservation is a concept that has been with us since at least the 1960s. Sustainable development is a concept that originated in the 1980s and has become one of the core agendas of our time. Although their roots are different, *conservation* and *sustainability* share common ground.

This book sets out to examine how these issues relate to each other in the context of historic cities. It aims to:

- identify weaknesses in current philosophy and practice in urban conservation;
- set out the relationship between successful architectural conservation and wider agendas of sustainability and cultural identity;
- summarise the communality of approach and practice that needs to be fostered and developed between a complex range of interrelated issues and disciplines;
- enhance both the perceived relevance of architectural conservation and its level of attainment;
- extend the achievement of the goals of sustainability in the context of historic cities; and
- highlight the opportunities for conservation and sustainability to work in a partnership of profound strength and mutual achievement.

Previous publications have focused on conservation and sustainability as though they are separate issues. Perceptions, however, are changing, new linkages are being forged, and this book seeks to contribute to this important process.

The starting point for this book is the United Kingdom. It draws from examples of theory and practice across Europe and elsewhere around the world.

Acknowledgements

The origins of this book date back to two post-graduate dissertations that I prepared at University in the early-1970s, one that related to Edinburgh, the other to countries across Western Europe. Over the intervening period the ideas and material that have gone into its preparation owe so much to so many people that it would be invidious to name them individually and risk omitting some of the most influential – including many whom I have never met and know only through their work. Suffice it to say:

First, that the support of colleagues, contemporaries and friends in cities as diverse as London, Bath, Cambridge, Derby and Edinburgh within the United Kingdom, and in cities across the length and breadth of continental East and West Europe and beyond, has been decisive.

Second, that my interest and perseverance in architectural and urban conservation and their relationships to modern architecture and town planning owe an enormous debt to the support of the late Sir Leslie Martin. My professor at Cambridge and a noted modernist in his own professional career, he had an inclusive approach to the challenges that budding architects should be prepared for in the future, and supported and encouraged me at a time when the topics were even less fashionable in the generality of the schools of architecture than they are today.

Definitions: Conservation and Sustainability

The context of this book is historic cities and the theory and practice of *conservation* and *sustainability* in relation to them.

In the wider, environmental sense, *conservation* and *sustainability* have parallel meanings and are frequently used interchangeably to express the need to manage the world's natural resources and the biosphere in order: first, to secure long-term harmony between man and nature; and second, to achieve continuous enhancement in the environment and in the conditions and quality of life for humans and other life forms.

It is in this broad sense that *sustainability* is used in this book.

Conservation, on the other hand, has a much narrower meaning when applied to historic cities. The principal root is architectural conservation, whose starting points include archaeology and the geocultural diversity and historical evolution of architectural styles, building materials and techniques. The secondary root of urban conservation is townscape and a morphological and aesthetic approach to the management of change in historic cities. Neither architectural conservation nor townscape is founded upon a preoccupation with sustainability. Both, however, have the potential to make a significant contribution to it.

Today in most western countries it is the mis-spent wealth in development which is the biggest agent of the destruction of historic cities, not physical decay.

Graeme Shankland

Chapter 1
Conservation: Background

Architectural conservation: beginnings and evolution

The history of modern architectural conservation has been traced to the confluence of Christianity and Humanism at the time of the Italian Renaissance, and to the recognition of classical antiquity both as an important epoch of the past and a springboard for cultural continuity and creativity. Monuments – whether ruined or otherwise – became prized for their inherent architectural and visual qualities as well as for their historical and educational value; in short, as predominantly today, for their architectural and historic interest (Figure 1.1).

By the eighteenth century – the Age of Reason (or Enlightenment) in Europe – advances in the sciences, coupled with an increasingly adventurous interest in the antiquities of Greece and Rome, led to the development of methodologies for verifying facts from primary sources and the founding of modern archaeology and art history. The concept of authenticity based on reliable information sources, which is a cornerstone of modern conservation philosophy and practice, is a product of that time.

The eighteenth century is also noted for the Picturesque Movement, inspired by romanticised paintings and engravings depicting the landscape and ruins of antiquity and by a growing interest in the medieval period. In England, this movement expressed itself in the landscaped park, and the protection, restoration and frequent construction of replica or false ruins for their picturesque value (Figure 1.2). Today, ruins – which in the United Kingdom are classified separately from historic buildings as ancient monuments – continue to be prized for their picturesque value, and there is a presumption by some against their being brought back into human occupation and practical use.

The eighteenth century also saw the beginnings of custodianship and systems of protection, together with the emergence of a sense of common ownership whereby important works of art and other authentic manifestations of a particular culture are seen as transcending national barriers. The collections of many of the great museums and art galleries of the world date

Figure 1.1 Urbino, Italy: the courtyard of the Ducal Palace (built c.1470–75; Luciano Laurana, architect). The Italian Renaissance was inspired by classical antiquity, which it saw as both a source and an inspiration for its arts and architecture.

from this time, and common ownership and responsibility – for both cultural and natural heritage – are expressed today primarily through the *World Heritage Convention*.

The nineteenth century in Europe saw the spread of the Romantic Movement, the emergence of nationalism, and the recognition of cultural diversity and pluralism. Increasing importance was attached to national, regional and local identity, and to the preservation of historic buildings, works of art and other expressions of individual geocultural identity.

Figure 1.2 Fountains Abbey, with its ruined twelfth-century Cistercian monastery set in an eighteenth-century landscaped parkland, represents the apogee of the Picturesque Movement in England. The earliest published use of *picturesque* has been traced to 1768, when it was defined as 'that kind of beauty which would look well in a picture'.

Over a time-span of more than five centuries, interest in historic buildings has expanded from its beginnings with the ruins of classical antiquity and the monuments of the early-Renaissance that they inspired to encompass the Romanesque and Gothic periods and, progressively over time, all architectural variants and styles up to and including our own. From an initial focus on major individual monuments and ensembles it has spread to include historic gardens, domestic architecture and the vernacular, the historic areas of cities, industrial archaeology, the Modern Movement and

Turning points in the history of architectural conservation: eighteenth to twentieth centuries

French Revolution

The former possessions of the king, the aristocracy and the church, initially the targets of destruction as symbols of former oppression, were soon recognised as testimony to the past achievements of the people who had formed the French nation. As such, the nation had responsibility to care for and protect them. Today, state intervention in one form or another is a characteristic of architectural conservation in most countries of the world.

Figure 1.3 St Albans Cathedral, England, was substantially remodelled in the name of restoration: 1860–78 by Sir George Gilbert Scott, architect; followed, 1878–83, by Lord Grimthorpe. Scott's contemporary, Eugène Emmanuel Viollet-le-Duc, defined restoration as follows: 'to restore a building is not to preserve it, to repair it, or to rebuild it: it is to reinstate it to a condition of completeness which may never have existed at any given point in time'.

Nineteenth century

The speculative, stylistic *restoration* (as it was termed) of Gothic churches and cathedrals, regardless of authenticity and historical layers, and of which the major proponents were the architects Sir George Gilbert Scott (1811–78) in England and Eugène Emmanuel Viollet-le-Duc (1814–79) in France, led to a strong anti-restoration movement in both countries (Figure 1.3). In England, this movement inspired the foundation of the Society for the Protection of Ancient Buildings (SPAB). The *SPAB Manifesto* of 1877, drafted by William Morris, is often cited as the formal basis for architectural conservation.

Second World War and its aftermath in Europe

The destruction of the historic hearts of numerous cities – from Plymouth in South-West England, through Rotterdam in the Netherlands, to Dresden in what became East Germany – set in train major schemes of reordering, reconstruction and redevelopment that affected cities across Europe. Generally, these embraced the theories of the Modern Movement in architecture and planning, and were extended to encompass cities that had been unaffected

Figure 1.4 Bath, England. Queen Square (built 1728–36; John Wood the elder, architect) was one of the earliest developments in the Georgian period (1714–1830). It was the first neoclassical development in Bath to unite a row of terraced houses behind a single Palladian-style palace facade; its central garden was laid out as a parterre, in the manner of a country mansion. At the time it was built, Queen Square was a quiet residential development outside the medieval city walls, without through traffic. By the early-1970s a series of highly destructive proposals for the city culminated in a major public and media campaign to *Save Bath*. A traffic and transport study prepared by Colin Buchanan and Partners in 1965 would have led to the demolition and reconstruction of buildings immediately behind the south side of Queen Square to allow a new cut-and-cover road to pass underneath, and to a new motorway-standard tunnel bored under land behind the north side (shown here).

by the War. The resultant destruction – especially in city centres across the length and breadth of England – was extensive. From the 1950s onwards, this destruction was combated by the emergence of activist groups and a string of well-orchestrated campaigns to save individual historic buildings and whole cities, of which one of the most influential was the campaign in the city of Bath (Figure 1.4).

European Architectural Heritage Year 1975

The year 1975 marked a turning point in the tide of post-Second World War destruction, especially in historic towns and cities; also of certain historic building types – in the United Kingdom, for example, of country houses. Led by the Council of Europe, European Architectural Heritage Year constituted a Europe-wide campaign of awareness-raising and action (Figure 1.5).

Figure 1.5 Brussels, Belgium: the Grand-Place in 1971. The elimination of parked cars and coaches in the Grand-Place was one of many key projects of European Architectural Heritage Year 1975 and inspired similar pedestrianisation schemes across Europe.

beyond; and from ruins through state-owned monuments to privately owned properties and multi-ownership settlements. Interest has also expanded geographically to include non-European cultures, including indigenous ones.

There has been a simultaneous expansion of study and awareness of traditional building materials, constructional techniques and craft skills, and increasing concern to secure their use in any and every intervention involving historic buildings: from repairs, through adaptations and alterations, to restorations. The retention of authenticity is an accepted – if not always observed – precept; at least, in the European cultural context.

Architectural conservation has evolved from the partly educational and inspirational, partly romantic and nostalgic preservation of individual buildings into a broad discipline supported by a number of key international governmental and non-governmental organisations, the subject of numerous charters, conventions, declarations and manifestos. Concurrently, especially by embracing inhabited settlements such as the historic areas of cities, it has confronted issues that extend substantially beyond those at its nucleus.

Nevertheless, architectural conservation remains rooted in its essentially European, Christian and monumental origins, underscored by a curatorial approach that is dominated by academics, archaeologists, specialist professionals and crafts, and with a protective, legislative basis that is contingent upon architectural and historic interest.

The language of architectural conservation

The practice of architectural conservation employs a number of key words that have taken on specific meanings, changed over time, been used synonymously, or been redefined from those in common usage – none of which is helpful to a wider appreciation of what conservation is or has the potential to achieve.

Heritage

First, the over-arching word of *heritage*. Etymologically, heritage is related to patrimony and signifies possessions and traditions that are inherited and passed on. The United Nations Educational, Scientific and Cultural Organisation (UNESCO) defines *heritage* broadly and well: 'heritage is our legacy from the past, what we live with today, and what we pass on to future generations'. In this definition heritage is neither limited in time nor restricted to material objects – whether they be historic remains, buildings, artefacts or whatever. Heritage is interpreted as the foundation of the present, the springboard for the future, with the present generation as its custodians and the creative link.

To many, however, heritage has a far more limited meaning, for example, 'the culture, property, and characteristics of past times'; or, 'today's perception of a pattern of events in the past'. As such, heritage has become a construct, a concept that relates only to history, that can be packaged for education and tourism, and that is perceived to be divorced from individual and community life today. One of the consequences of this construct is to limit perceptions of the purpose of architectural conservation: first, for the preservation of historical evidence; and second, to provide fuel for the heritage industry. The naming and public profile of government agencies such as English Heritage and Historic Scotland reinforce this construct.

Preservation, restoration and conservation

Second, the triad of *preservation*, *restoration* and *conservation*. Even within the architectural conservation fraternity these words continue to be used interchangeably; not least because *conservation*, the most fashionable word in the English-speaking world today, does not translate well into other tongues – where its usage is mostly confined to the care of works of art and other objects in museum environments.

The concept of *restoration*, derided in some quarters since the mid-nineteenth century, is still favoured by many as a word that communicates positively and universally. Some practitioners in the United Kingdom still use *preservation*, as that was the fashionable term until it was superseded by *conservation* in the 1980s.

Many of the older conservation charters use any pair or all three of these words without attributing discrete meanings, and the inter-governmental organisation whose present-day acronym is ICCROM – which stands for the International Centre for Conservation in Rome – functions under the official title of the International Centre for the Study of the Preservation and Restoration of Cultural Property, thereby succeeding to employ all three words at once, and in a manner that implies that *preservation* + *restoration* = *conservation*.

This formula is somewhat borne out by the *Burra Charter* (first published in 1979, most recently revised in 1999) which offers the definitions that are favoured by conservation purists – but which do not necessarily communicate well to a wider audience:

> *Preservation* means maintaining the fabric of a place in its existing state and retarding deterioration.
>
> *Restoration* means returning the existing fabric of a place to a known earlier state by removing accretions or by reassembling existing components without the introduction of new material.
>
> *Conservation* means all the processes of looking after a place so as to retain its cultural significance.

Unfortunately, the definition here of *restoration* is curatorial, represents an enhanced form of *preservation*, and has little relevance in the case of historic buildings in use where new material may be introduced for a whole series of practical and aesthetic reasons, not least to make a building wind and watertight and for the recovery of architectural integrity. In order to resolve the problem created by this definition, the *Burra Charter* adds to the triad, as follows:

> *Reconstruction* means returning a place to a known earlier state and is distinguished from restoration by the introduction of new material into the fabric.

Generally – Scott and Viollet-le-Duc excepted – this definition would be taken to relate to *restoration*; *reconstruction* is understood to mean taking down followed by rebuilding.

Sir Bernard Feilden offers a more coherent explanation of *restoration*: 'The object of restoration is to revive the original concept or legibility of the [building]'. As a conservation architect myself, this is how I also understand and use *restoration* (Figures 1.6 and 1.7).

Authenticity

Third, *authenticity*, a word that is used on the international stage but is unhelpfully and ambiguously subsumed into *character* and *appearance* in United Kingdom protective legislation and practice. Authenticity is defined in an ICCROM publication (essentially in a European context) as: 'materially *original* or *genuine* as it was constructed and as it has aged and weathered in time'. Meanwhile, whole sections of the building industry thrive on the ambiguity of *character* and *appearance*, whether they are marketing period homes or plastic, look-alike, windows and doors.

Sciennes Hill House: *before restoration*.

Sciennes Hill House: *after restoration*.

Figures 1.6 and 1.7 Edinburgh, Scotland: restoration of Sciennes Hill House. The original mansion house, built in 1741, was partially demolished in 1868 and incorporated into a street of tenement housing. Archaeological and documentary evidence enabled the architectural integrity of the principal elevation to be recovered (restored 1989; Dennis Rodwell, architect).

Conservation charters

The philosophy and practice of both architectural and urban conservation are informed by an ever-increasing number of national and international charters and declarations that date back to the nineteenth century. There is a sense in which these documents constitute an essentially intellectual exercise. Certainly, they reflect a predominantly European tradition of philosophy and practice.

Each of these documents is a product of its time, place and authorship. Certain of them, for example the *SPAB Manifesto* of 1877, are overtly reactive to a particular situation – in this instance, one that is essentially in the past – and are purposefully proscriptive. Others are proactive and positive. Despite many inconsistencies and contradictions, a common thread is the practical one of seeking to position conservation with an identifiable ethic and methodology and, increasingly over time, to reconcile it with other interests.

Common to all of the charters is their focus on the protection of selected buildings or groupings that are either explicitly or implicitly characterised as monuments – *monument* being a word whose origins relate to memorial, inscription and other cultural expressions that are perceived as documents from the past and essentially permanent, and which transmit messages or values from one generation to the next. Later charters expand the concept of values beyond the purely cultural into the social and the economic. Common also – as the underlying guiding principle of modern conservation – is the concept of authenticity, and its protection and retention in all interventions.

Both the 1931 *Athens Charter* and its direct successor the 1964 *Venice Charter* support the use of modern materials and techniques. In consequence, reinforced concrete – one of the favoured constructional techniques of the Modern Movement in architecture – came to be used in many post-Second World War restoration projects throughout Europe. It continues to be used across parts of the former Eastern Bloc. Today, more generally, the incompatibility of modern materials and techniques with the environmental performance and structural integrity of traditionally built historic buildings is recognised and steps are taken to avoid their use (Figure 1.8).

A notable inconsistency between charters occurs over the parameters that should be applied to the design of new buildings in the surroundings of historic monuments and within historic areas. The 1931 *Athens Charter* urges respect; the 1933 *Charte d'Athènes* condemns the reproduction of historical styles; the 1964 *Venice Charter* insists that new structures should be distinct and contemporary; and the 1975 *European Charter* promotes the use of traditional materials. In none of these is the Renaissance concept of creative continuity expressed, nor the overtly derivative nature of all architectural design until at least the onset of the Modern Movement – historical derivation

Figure 1.8 Dubrovnik, Croatia, lies in a region that is subject to regular seismic activity. The most recent major earthquake took place in 1979 and damaged over a hundred monuments in the walled city and its immediate surrounds. Encouraged by experts schooled in the 1964 Venice Charter, reinforced concrete was used in the restoration of monuments in and close to the main street, the Stradun (shown here); in particular, the Rector's Palace. The rigidity that this introduced to their structures ill-equipped them for subsequent tremors. The Bishop's Palace has now been restored using enhanced traditional methods designed to allow its structure to flex.

being a formative characteristic of the architecture of the towns and cities that pre-date it. Only in the 1987 *Washington Charter* is the potential for contemporary elements to contribute to the enrichment of a historic area expressed – with the proviso that they be 'in harmony'. These inconsistencies reflect a variety of attempts to reconcile the philosophy and practice of conservation with the education and practice of architecture today. It is a debate that remains unresolved.

Key charters and their contexts

1877 SPAB Manifesto

The *SPAB Manifesto* reacted forcibly and uncompromisingly to the mid-nineteenth-century fashion for the stylistic remodelling of Gothic monuments – without respect for historical layers and in direct confrontation with the concept of authenticity. The Manifesto is focused on 'ancient monuments of art' – whose qualities may include historical and picturesque. It berates 'restoration' and exhorts 'protection'.

The *SPAB Manifesto* laid down two principles: first, that of minimum intervention (expressed as 'to stave off decay by daily care'); second, that where a monument is no longer considered suitable for use without being altered or enlarged it should be taken out of use and preserved as it stands.

1931 Athens Charter

It is both confusing and significant that there are two Athens Charters from the 1930s. The context for both was the Modern Movement in architecture and planning and the constructional techniques and design concepts that were favoured by it.

The 1931 *Athens Charter* was the first document to set out the scientific principles for the preservation and restoration of historic monuments at an international level. Focused on historic sites which are subject to strict custodial protection, it supported the use of modern materials and techniques in restoration work, favoured continuity of appropriate use, recommended respect for the surroundings of monuments including in the design of new buildings, and urged increasing international cooperation.

By contrast, while the 1933 *Charte d'Athènes* – one of the seminal manifestos of international modernism – recognised the protection of individual monuments and urban ensembles, it condemned any attempt at aesthetic assimilation through the use of historical styles for new structures in historic areas.

1964 Venice Charter

The *Venice Charter* represented a revision of the 1931 *Athens Charter*. It again supported the use of modern techniques, emphasised the importance of authenticity based on material and documentary evidence, and extended the concept of historic monuments to include urban and rural settings.

The *Venice Charter* stated that the aim of conservation and restoration work is to safeguard monuments both as works of art and as historical evidence. Accordingly, where components are replaced they should be integrated harmoniously but be distinguishable, and any additions to a monument should be distinct and contemporary.

This charter was adopted as the principal doctrinal document of the International Council on Monuments and Sites (ICOMOS) when it was founded the following year, 1965, and continues to be cited as the baseline document for international conservation philosophy and practice today.

1975 European Charter of the Architectural Heritage

The *European Charter* was adopted by the Council of Europe at the conclusion of European Architectural Heritage Year, and was complemented by the *Declaration of Amsterdam* outlining the basis for implementing it (Figure 1.9).

Figure 1.9 Amsterdam, Netherlands. European Architectural Heritage Year 1975 concluded with the Congress of Amsterdam. The *European Charter* coupled with the *Declaration of Amsterdam* were amongst the first international documents to promote integrated urban conservation – in which the architectural heritage is placed on an equal footing with other factors in the town and regional planning process.

The *European Charter* extended the concept of historic monuments to include urban and rural areas (as opposed simply to their settings), and emphasised the importance of passing the architectural heritage on to future generations in its authentic state. It recognised social and economic values in addition to cultural ones; also, that the future of this heritage depends largely upon its integration into the context of people's lives and the weight attached to it within the framework of general planning policy.

The *European Charter* promoted the concept of integrated conservation, in which priority is attached to retaining functional and social diversity in historic areas and to resisting the demands of motor traffic and the pressures of land and property speculation. It acknowledges modern architecture in historic areas – but on condition that the existing context, proportions, forms and scale are respected and traditional materials are used.

1987 *Washington Charter*

The *Washington Charter* complemented the 1964 *Venice Charter*. It paraphrased the conservation of historic towns and urban areas as: 'those steps necessary for the protection, conservation and restoration of such towns and areas as well as their development and harmonious adaptation to contemporary life'.

The *Washington Charter* states that urban conservation should be integral with socio-economic development and urban and regional planning policies at all levels. It represents the multi-disciplinary nature of urban conservation, emphasises the importance of active participation by residents – whom it sees as the primary stakeholders – and insists that the improvement of housing be regarded as a primary objective.

This ICOMOS charter summarises the important qualities that should be preserved: urban layout and grain; relationships between buildings and green and open spaces; relationships between a historic area or town and its surrounding man-made and natural settings; the diversity of functions as accumulated over time; and the exterior and interior appearance of buildings – from scale, through style and materials, to colour and decoration.

1999 Burra Charter (revision)

The stated aim of the *Burra Charter* is to provide 'guidance for the conservation and management of places of cultural significance'. As such, it is not exclusive to historic buildings or urban areas, and encompasses, for example, landscapes modified by human activity. Its guiding principles are closely related to the various charters that preceded the publication of its first edition in 1979 and that coincide with the twenty-year period of its several revisions.

The over-arching principle in the *Burra Charter* is the importance of understanding and safeguarding significance, including through the informed unravelling of historic layers, in ways that encapsulate a place's aesthetic, historic, scientific and spiritual values: from the past, in the present, and for the future (Figure 1.10).

The *Burra Charter* adopts a curatorial and scientific approach – one that distinguishes between old and new fabric and permits alterations on condition that they are considered both temporary and reversible. It also urges continuity of historical uses wherever possible. The *Burra Charter* concludes with an important message: 'the best conservation often involves the least work and can be inexpensive'.

Figure 1.10 Sydney, Australia. The precepts of the *Burra Charter* underscore present-day architectural conservation at an international level, especially on historic monuments. Statements of significance and conservation plans, such as for the Opera House (built 1957–73; Jørn Utzon, design architect), are products of this charter.

Urban conservation: museological beginnings

Marais quarter, Paris, France

France is credited with initiating the first major projects of urban conservation in Europe, following the enactment of the *Loi Malraux* in 1962. This established the legal and financial basis for *secteurs sauvegardés* (protected areas).

The designation of a *secteur sauvegardé* leads to the preparation of a comprehensive plan, one that encompasses town planning, architecture and detailed historic building conservation issues. One of the first of these plans to be prepared and implemented was for the Marais, until the early-eighteenth century the fashionable aristocratic quarter of the French capital. The designated area of the *secteur* covers 126 hectares, and incorporates the major part of the third and fourth *arrondissements* (districts).

During the nineteenth century the Marais became an artisan quarter, and the former *hôtels particuliers* – the town mansion houses of the rich – were taken over and subdivided into workshops and apartments, their courtyards often built over to form warehouses. By the end of the Second World War and with a population of well over 80,000, the Marais had become seriously dilapidated. Surveys completed by 1960 indicated that around sixty per cent of its dwellings lacked toilets; thirty per cent, running water; and fifteen per cent, electricity (Figure 1.11).

Figure 1.11 Marais quarter, Paris, France. The initial plan for the *secteur sauvegardé* was highly interventionist. It was aimed at restoring the quarter to its early-eighteenth-century architectural splendour. This view of the centrepiece of the Marais, the Place des Vosges (built 1605–12; originally known as the Place Royale), was taken in 1960, two years before the enactment of the *Loi Malraux*.

The original plan for the *secteur sauvegardé* was a highly interventionist one aimed at the restoration of the entire quarter to its former glory, guided by the footprint of a plan that dated from 1739. The *secteur sauvegardé* plan anticipated the restoration of all of the historic buildings externally and internally, the opening up of the spaces between buildings and within courtyards that had been built over, and the recreation of the gardens. The key historic buildings were the *hôtels* (Figures 1.12 and 1.13); additionally, there were many hundreds of more modest houses and apartment buildings.

At a conference held in Edinburgh in 1970, François Sorlin – then Inspector General for Sites and Historic Monuments at the French Ministry of Cultural Affairs – stated the position as set out in the conservation plan for the Marais as follows: 'the only solution for the revitalization of the 300 large residences in the Marais is to use them for embassies or head offices of large companies'. Museums and central and local government offices were also considered compatible uses (Figures 1.14 and 1.15).

A few of the buildings in the Marais date back to the eleventh century; there are some half-timbered houses from the fifteenth century, and several buildings from the nineteenth century. Predominantly, the architecture of the area dates from the seventeenth and eighteenth centuries. As the primary

Hôtel le Rebours: *street elevation*.   Hôtel le Rebours: *the courtyard*.

Figures 1.12 and 1.13 Marais quarter, Paris, France. The seventeenth-century Hôtel le Rebours in the Rue Saint-Merri awaiting restoration (1990).

objective for the Marais was to restore the quarter up to a notional ideal date sometime towards the mid-1700s, the *secteur* plan anticipated – in addition to the removal of all later accretions to the buildings identified as

Hôtel de Sully: *street elevation.*

Hôtel de Sully: *garden elevation.*

Figures 1.14 and 1.15 Marais quarter, Paris, France. The Hôtel de Sully in the Rue Saint-Antoine (begun in 1625) was one of the first of the great town mansion houses to be restored in the 1960s. Today it is the headquarters of the *Caisse Nationale des Monuments Historiques de France.*

monuments – the demolition of nineteenth- and twentieth-century structures, and the use of earlier architectural styles for their replacements and for any infill developments. There was a clear resonance here with the work and writings of Eugène Emmanuel Viollet-le-Duc a century earlier.

Initially, the plans for all *secteurs sauvegardés* were foreseen as rigid blueprints. By the mid-1970s, the apparently inflexible and overridingly architectural, historical and museological approach to the Marais softened.

First, there were not 300 end-users for immaculately restored *hôtels*, so other options needed to be explored. Second, the initial plan anticipated a considerable movement of population from the Marais to the suburbs (the French word for suburb, significantly, is *banlieue* – or 'place of banishment'). Riots had taken place in the part of the Avignon *secteur sauvegardé* known as *La Balance* following the forced rehousing of its population elsewhere in the city as part of the regeneration programme there, and the politics had to be rethought at a national level (Figures 1.16 and 1.17). Third, the use of architectural pastiche was contested.

Avignon: *restored eighteenth century houses.*   Avignon: *new block of apartments.*

Figures 1.16 and 1.17 Avignon, France. It is a coincidence, but in France one of the turning points in the implementation of the *secteur sauvegardé* programme was the destruction in Avignon of buildings in the quarter known as *La Balance*, whilst at Bath in England, one of the turning points in the *Save Bath* campaign was the destruction of Ballance Street, a street of Georgian artisan housing on Lansdown Hill. Shown here at *La Balance*, Avignon, shortly after their completion in 1971, are the restoration of a group of eighteenth-century houses and the development of a new block of upmarket apartments opposite.

Ancient Reserve, Plovdiv, Bulgaria

A parallel example of a museological approach to urban conservation has been pursued without modification in Plovdiv, Bulgaria's second city (population $c.500,000$). With a history that dates back over 6,000 years, Plovdiv is one of Europe's oldest cities.

The historic core boasts over 200 town mansions dating from the mid-nineteenth century and built in the national revival style – also known as Bulgarian Renaissance (Figure 1.18). Of these, a significant proportion are derelict, underused, or in need of extensive programmes of conservation and maintenance. The historic area covers 35 hectares and, following the relocation of most of its residents to other parts of the city several decades ago, it has a residual population of around 1,500. The area is designated a cultural and tourist zone and is described as the *Ancient Reserve*. To date, there has been a very limited perception of appropriate uses for these town houses, the over-riding policy being to restore and apply cultural uses to them, uses that make them accessible to the public for higher education or as visitors but detach them from the everyday life of the majority of citizens. Adaptation to other uses, even their reversion to the residential uses for which they were constructed, is not encouraged.

Figure 1.18 Plovdiv, Bulgaria. The programme of restoration in the *Ancient Reserve* was commenced in the 1950s. The Geordiady House (built 1848) was restored to house the Museum of National Revival.

Figure 1.19 Plovdiv, Bulgaria. Around half of the more than 200 monuments in the *Ancient Reserve* are underused, in poor condition or derelict; some are in ruins. Detaching this historic area from the everyday life of the modern city has seriously limited the options for using these monuments, and therefore the investment to restore them.

There are only so many art galleries, museums, libraries, and institutes that any city can support. The *Ancient Reserve* has taken on the aspect of an open-air museum, with its associated complement of souvenir shops and stalls, and the city is struggling to find either investment or uses for the many derelict houses (Figure 1.19).

Townscape: the concept

Townscape is the cornerstone of urban design, a discipline that runs parallel to architectural conservation in the urban context and forms a bridge between architecture and town planning.

In his seminal book of 1961, *Townscape*, Gordon Cullen defines *townscape* as the art of coherent three-dimensional composition, one in which the individual components of any urban landscape – the buildings, the spaces and enclosures, the connections and closures, the vistas and views – knit together to form a set of relationships that are at one and the same time harmonious and contrasting, static and changing, and whose combined impact determines the physical sense of place and identity.

The characteristics that are described as giving individual personality to an urban environment include the design of the buildings – their scale, style, texture and colour; contrasts such as grandeur and intimacy, blandness and intricacy; hard and soft landscaping; street furniture; the calligraphy used in signage and advertising; and reactions such as anticipation and surprise as one moves about.

Cullen appreciates that older towns and cities have been created over time, usually organically, and that they embrace different periods and architectural styles. He encapsulates the essence of good urban design as: 'the agreement to differ within a recognised tolerance of behaviour'.

Cullen's treatise coincided with the publication of *The Image of the City*, in which Kevin Lynch analysed how people read and relate to visual form as they move about cities by framing mental images based on five elements: paths, edges, districts, nodes and landmarks. First published in America in 1970, it has made a significant contribution to city planning and civic design at the scale of the modern city.

Chapter 1: digest

Architectural conservation is rooted in an essentially European, Christian and monumental tradition that prioritises a scientific approach to evidence. From the starting point of architectural and historic interest, influences have included the Picturesque and Romantic Movements and, progressively from the late-eighteenth century onwards, reaction to revolutions, to wars, and to contentious interventions affecting individual building types and entire cities.

Over a period of several centuries, architectural conservation has developed from an elitist interest in key monuments of major stylistic periods to a broad discipline that recognises values in a spectrum of building types and epochs, in the range of scales from the rural vernacular to the historic city, and attaches importance to geocultural diversity.

The subject of numerous charters and related documents, architectural conservation is nevertheless a specialism that is contingent upon value judgements related to architectural and historic interest. The sequence of charters follows a logical progression by place and time, and discloses a variety of attempts to position and reconcile architectural conservation with other interests: within the architectural profession, with the Modern Movement and its successors; and within the town planning context, with a broad range of outside forces and skills. The *European* and *Washington Charters* are important in paving the way for a multi-disciplinary approach to urban conservation.

The scientific basis is underscored by a curatorial approach to practical conservation that prioritises the concept of *authenticity* together with the use of traditional materials, constructional techniques and craft skills. The *Burra Charter* presents important messages about understanding significance and minimising the extent and cost of works. Architectural conservation is, however, handicapped by confusion and ambiguity in key elements of its vocabulary: *heritage, preservation, conservation, restoration* and, not least, *authenticity*.

Early essays in urban conservation – notably under the programme of *secteurs sauvegardés* across France – disclosed a museological approach that was divorced from its wider socio-economic and town planning context, and was accordingly reviewed. The origins of the concept of *townscape* are, likewise, unrelated to the wider socio-economic and town-planning context, but the analogy between good urban design and 'the agreement to differ within a recognised tolerance of behaviour' is a valuable one.

Chapter 2
Urban Planning Context

The pre-industrial city

The generality of European cities of the pre-industrial era – whether they were planned at a specific date or developed organically over time – held certain functions and elements in common.

They were centres of power, trade, and social and cultural interaction. They were clearly defined and compact – whether for defensive or administrative reasons – and relatively densely populated. They had few major buildings: a palace or castle; religious buildings; a guild or town hall; and an exchange. They were diffused with craft industries and traders – often integral with residential accommodation – and they housed mixed communities: always socially, and often, especially further east in Europe, by religion and ethnic origin. The market place was the focal point for trade, sited strategically within a city so as to serve its citizens best and to entice travellers passing through.

Such cities related to their topography and enjoyed a balanced relationship to their locality – whether to the sea, rivers, forests or fields – and to the outlying communities that depended on them. Their sense of place and harmony was enhanced by the limited range of local materials and craft skills used in their construction, sometimes reinforced by strict building codes (Figure 2.1). Their scale was essentially human, and they functioned socio-economically according to a commonly shared understanding of what constitutes urban life as compared to rural life.

As with architectural conservation, several of the aspects of townscape – that is, of the art of coherent three-dimensional composition – may be traced back at least to the Italian Renaissance: order; views and vistas; relationships between buildings, public and private spaces; and contrast between public ostentation and private intimacy.

Additionally, pre-industrial cities were characterised by the hierarchical organisation of activities according to type and scale, but always in close proximity, and primary and secondary circulation by horse-drawn vehicle or by foot – the latter frequently by passages that interconnect streets, sometimes passing through courtyards (Figures 2.2 and 2.3).

Figure 2.1 Dubrovnik, Croatia. Strict urban planning regulations were first introduced in 1272. They ensured the overall harmony of the Gothic, Renaissance and Baroque architecture of this city-state through the control of building heights and materials. Today, they also control paint colours, advertising, and the retention of historic architectural features.

Figure 2.2 Zamość, Poland. The Renaissance concept of order in urban planning was often expressed through geometric diagrams based on the square, the circle, and the polygon. Founded in 1580, Zamość was laid out and built by Bernardo Morando, an architect originally from Padua. It is the epitome of an ideal late-Renaissance (or Mannerist) city-fortress, complete with its princely residence, town hall, market square, churches, a grid pattern of arcaded streets, defensive walls and bastions.

Figure 2.3 Zamość, Poland. In a traditional historic city all external and internal space is fully utilised – whether for commercial, social or cultural interaction, workspace or habitation. Residential is normally the majority use of floorspace; there are no vacant upper floors, no disused or derelict backlands behind and between streets. Here at Zamość, some of the inner residential courtyards are on pedestrian through routes, others are cul-de-sacs and more private; all provide a safe environment for children and adults of all ages.

The mainstream of modern town planning

The roots of a utopian vision

The formative influence in modern Western town planning lies not with the pre-industrial city but with the pioneering experience of the Industrial Revolution in Britain from the late-eighteenth century onwards. The new industrial city coincided with agrarian reform. It was characterised by the concentration of factories in cities, migration from the countryside, and the use of coal as the primary source of power and heat for industry and the home. The evolution of the industrial city through the nineteenth and into the twentieth centuries ran parallel with developments in the mechanical transportation of goods and people: first, with railways supplanting the canal system; then, with railways facing increasing competition from transport by road.

The harsh conditions of the Victorian city in Britain were epitomised by acute urban pollution, high rates of infant mortality, and successive cholera and typhoid outbreaks. Graphic illustrations of this harshness reached international audiences through the literary imagery of Charles Dickens (1812–70) and the visual imagery of Gustave Doré (1832–83).

Concerns for public health and working-class housing conditions, coupled with strong undertones of paternalistic socialism and romanticism, developed into an uncompromising anti-urban movement that idealised an

Figure 2.4 Rottingdean, England: early-twentieth-century cottages in the late-medieval Tudor style. The anti-urban movement idealised a quasi-rustic way of life based on the unitary family with its house and garden.

alternative, quasi-rustic way of life based on the unitary family with its house and garden (Figure 2.4). Both the author and artist John Ruskin (1819–1900) and the writer and designer William Morris (1834–96; author of the *SPAB Manifesto* of 1877 – see page 12) played important roles in this movement.

This anti-urbanism evolved into a new utopian vision, one that denied the pre-industrial concept of urbanity: the built forms and relationships that reflected it; and the traditional mixity and mingling of socio-economic activity that characterised it at all levels. Town planning in the United Kingdom especially has yet to recover from this anti-urban legacy, a legacy that inevitably places the theory and practice of planning in confrontation with the buildings and infrastructure of historic cities – in varying degree according to any individual city's history, evolution, and physical characteristics.

The roots of the new utopian vision are to be found in the paternalistic, philanthropic and moralist factory villages of the Derwent Valley, New Lanark and Saltaire, their extension through the model industrial villages of the English Midlands and North-West, and their parallels in North America and continental Europe, such as the *cités ouvrières* (social housing) and the *familistères* (cooperatives) of France (Figure 2.5).

Figure 2.5 Darley Abbey, England, is the most complete of the surviving late-eighteenth to early-nineteenth-century factory villages in the Derwent Valley. With its compactly spaced but generously sized workers' housing interspersed with allotments, separated by the river from the cotton mills that provided the employment, it formed part of the inspiration for the Garden City movement. Brick Row in Darley Abbey (built 1797–1800) was designed to the Georgian urban model of a terrace of houses. Later developments elsewhere adopted quasi-rural models, precursors of suburban housing types – as shown in Figure 2.4.

The Garden City and the Modern Movement

By mutation in the hands of Sir Ebenezer Howard (1850–1928), the utopian vision was expressed coherently and to an international audience as the Garden City, arguably the single most important formative concept of modern town planning. Howard's simplistic geometric diagram for a Garden City was first published in 1898. It was conceived for a population of 30,000; to house larger numbers, Howard proposed they should be developed in clusters. The diagram was based on a circle with concentric rings, with both concentric and radial routes of primary and secondary circulation. Unlike, however, its parallels with the geometrically planned ideal city of the Italian Renaissance, which was a three-dimensional artistic vision related to architecture, public spaces and vistas, the Garden City was a simplistic, two-dimensional, quasi-sociological concept that had one over-riding objective: the allocation of land into separate zones for housing, recreation and industry, interconnected by a hierarchical pattern of circulation, and surrounded by agricultural land.

In the United Kingdom, Howard's concept led to the founding of Letchworth Garden City in 1903, Welwyn Garden City in 1919, the passing of the first Green Belt Act in 1938 and the New Towns Act in 1946.

The First World War devastated large parts of Belgium and Eastern France. Coupled to this, cities across Europe had inherited from the nineteenth century a legacy of ill-maintained and ill-serviced housing.

Leaders of the Modern Movement in architecture and planning perceived the principle of separation of uses interconnected by mechanised transport as the answer to the problems of the industrial city and metropolis. This was the primary focus of the 1933 *Charte d'Athènes* (see also page 12).

Echoing Sir Ebenezer Howard, this 1933 Charter set out the four functions of the city: dwelling, recreation, work and transportation. It anticipated the large-scale reconstruction of historic cities according to this simplistic model, including through the demolition of substandard housing, which it describes as 'insanitary slums'. These were to be replaced by open space, and new residential quarters were to be constructed to lower housing densities in other parts of the city. The city was also to be reordered to satisfy the needs of mechanised transport, which by now included the motor car.

Charles-Edouard Jeanneret (1887–1965), known as Le Corbusier, the Swiss-born architect, urban planner and adept promoter of the modernist vision, had already presented his proposal for the rebuilding of Paris. His *Plan Voisin* of 1925 showed the demolition of the entire Marais quarter, which he described as a particularly antiquated and unhealthy part of Paris, and its reconstruction as the new commercial neighbourhood with eighteen skyscrapers together with the rebuilding of a separate residential neighbourhood to its west (Figure 2.6).

As a town planner, it is easy to dismiss Le Corbusier as a megalomaniac fantasist whose ideas have been discredited, superseded, and are of no relevance today. But this is oversimplistic, and ignores their lingering, pervasive influence. Le Corbusier, his contemporaries and followers have inspired successive generations of architects and planners throughout the world (Figure 2.7).

Figure 2.6 Paris, France. View north-westwards from the tower of Notre Dame today, with its skyline essentially unaltered since the nineteenth century. Le Corbusier's *Plan Voisin* of 1925 would have led to the substantial rebuilding of Paris, with the exception of a selected few key monuments or relics from them.

Figure 2.7 Ronchamp, France. As an architect, especially, Le Corbusier continues to be revered as a creative genius. The chapel of Notre-Dame-du-Haut (built 1953–55) attracts visitors from all over the world.

Especially in Britain in the pre- and post-Second World War periods, adherents to the concept of the Garden City and the ideas of the Modern Movement set in train a series of destructive ideas that had a major impact on historic cities of all ages and sizes. Slum clearance became a core part of national housing policy; comprehensive redevelopment programmes aimed at focusing a limited number of uses in city centres – primarily offices and shops – became a central platform of national planning policy; and the challenge of reordering cities to accommodate the motor car was welcomed by the road transport lobby and the car manufacturing industry.

Albeit moderated and reformulated, elements of these destructive ideas remain with us today. In Europe, the United Kingdom especially has yet to recover the coherent, embracing, pre-industrial sense of what a city is and how it functions. In short, how to live wholly constructively and positively with the surviving historical inheritance of our towns and cities, including the industrial and metropolitan cities of the nineteenth century.

Alternative visions

From the latter part of the nineteenth century through to the mid-twentieth century, two important and discrete strands of thinking ran parallel to the mainstream of modern Western town planning. They operated at both a theoretical and a practical level and their leaders, Sir Patrick Geddes based in

Scotland and Gustavo Giovannoni in Italy, were contemporaries, respectively, of Sir Ebenezer Howard and Le Corbusier.

Both were polymaths – creative thinkers and rationalists rather than idealists. Both, by aptitude and application, were primarily educationalists; although they wrote and published widely, as teachers by instinct they relied more on the spoken word (also, in Geddes' case, a prolific correspondence) to explore, debate and develop their ideas. Accordingly, their principal published legacies lack order and cohesion; they speak of two charismatic individuals who were asking and seeking answers to questions as much as actually finding them. It has largely been left to others to attempt to consolidate, edit and express their conclusions into coherent, transmissible philosophies.

Sir Patrick Geddes

Sir Patrick Geddes (1854–1932), biologist, botanist, sociologist, town planner and property developer, studied the theory of evolution under Thomas Huxley (1825–95), a foremost follower of Charles Darwin (1809–82). In the aftermath of the social and environmental upheavals of the Industrial Revolution in Britain, Geddes pioneered a sociological approach to the study and practice of town planning, which he taught and applied extensively.

Evolution is the central theme throughout Geddes' life work. He defined the city as an ecosystem, one that needs to be understood as a living organism that is subject to cycles of birth, growth, blossoming, decline and decay, followed again by rebirth. Understanding the forces at work, anticipating the evolving functional demands of human society in the full complexity and interrelationship of their environmental, social and cultural aspects, enables control of the degenerative tendencies and continuous enhancement of the quality of life.

Geddes held passionately to the view that there is a direct evolutionary linkage between social and cultural development and place, and that the roots of a society's culture, including the heritage of its built environment, are the essential foundation for the achievement of citizens' creative potential, individually and collectively.

Geddes recognised that each city is unique, that its cultural evolution depended on its specific qualities of place and people, and expounded the importance of cultural identity and diversity. Emphasising the importance of culture in the life of cities, he also insisted upon the need to offset the centralising powers of a metropolis by stimulating cultural life in provincial cities – a view that was shared in France after the Second World War and enthusiastically taken up in the 1960s by André Malraux, France's first Minister of Culture, after whom the *Loi Malraux* is named.

Geddes expressed the interdisciplinary nature of town planning beyond the design-led professions, relating architecture to historical context – of place, people and cultural traditions – and spatial form to social processes. Planning and architecture were not therefore just about ordering the

Sir Patrick Geddes and conservation surgery in the Old Town of Edinburgh

> We now start with the idea that cities are fundamentally to be preserved and lived in; not freely destroyed, to be driven through, and speculated upon. (Geddes, writing in 1919)

In his early years, in the third quarter of the nineteenth century, Geddes witnessed major transformations in parts of the Old Town of Edinburgh. The medieval heart of the Scottish capital had fallen into serious disrepair and an extensive programme of improvement works was undertaken, comprising demolitions, the widening and formation of new roads, and much rebuilding (Figure 2.8). These works were concentrated in the area of the High Street, mid-way up the Royal Mile from the Palace of Holyroodhouse to the Castle, and had no impact on the appalling housing conditions downhill in the Canongate and uphill in the Lawnmarket.

Geddes sought to break the cycle of deprivation and decay in these less-favoured areas by applying his analogy between a city and an ecosystem and combining it with his belief in the importance of close proximity between job opportunities and place of residence – especially of low paid workers – and his thesis of the importance of culture in the life of a city. On this premiss, the Old Town and its citizens would flourish together.

To achieve this he projected a vision based on reviving the intellectual environment of cultural excellence in the mid-eighteenth century – the Age of the Enlightenment – when, as one

Figure 2.8 Old Town, Edinburgh, Scotland. The rebuilding of St Mary's Street in the 1870s formed part of the extensive programme of improvements that were carried out at that time in the area of the High Street. By the 1960s, in negation of Sir Patrick Geddes' ideas, this area had once again become blighted – this time by proposals for an urban highway – and a third of the apartments in St Mary's Street had fallen vacant. Between 1931 and 1961 the population of the Old Town of Edinburgh declined from 20,000 to 3,000. The road proposals were eventually abandoned and the whole street became the subject of another major programme, this time of upgrading and restoration. (Illustrated here: nos 6 to 14 St Mary's Street. Restored 1983–84; Dennis Rodwell architect.)

visitor was reported as saying: 'Here I stand at what is called the Cross of Edinburgh, and can, in a few minutes, take fifty men of genius and learning by the hand'.

An important component of this regenerative vision was to encourage students of Edinburgh University to live in the historic environment of the Old Town. In the 1880s and 1890s, Geddes adopted a hands-on approach to putting this vision into practice, undertaking directly and inspiring others to undertake a series of property developments in tenements in the area of the Lawnmarket, primarily as halls of residence (Figures 2.9 and 2.10).

Geddes pioneered what he termed *conservation surgery*, namely, the retention of all structures capable of renovation, improving and restoring both them and the courtyards between, and complementing them with the careful insertion of new buildings.

> There are finer architects than I, and bolder planners too: but *none so economical*.
> (Geddes, writing in 1922)

Old Town, Edinburgh, tenements: *rear, courtyard elevation.*

Old Town, Edinburgh, tenements: *front, street elevation.*

Figures 2.9 and 2.10 Old Town, Edinburgh, Scotland. Dating from the seventeenth century, the renovation and partial reconstruction of these adjoining tenements in the Lawnmarket formed part of Sir Patrick Geddes' vision for the regeneration of the Old Town. (Remodelled in the early-1890s; Stewart Henbest Capper, architect. Restored 1983–84; Dennis Rodwell, architect.)

physical environment; they were an essential component of social and cultural evolution and needed to be in harmony with it. Geddes' best-known published work, *Evolution in Cities*, is essentially a polemic on urban civilisation and citizenship.

By his diverse activities as theorist, practitioner and propagandist, Geddes sought to expand upon the relationship that existed in earlier times between the pre-industrial city and its immediate locality to encompass its region and, by extension, beyond: to the global relationship between cities and the world's resources. Geddes recognised that the well-being of the human species depended on achieving a new post-industrial equilibrium between people and their natural environment.

Viewed by some as a maverick, too flighty to be intellectually rigorous, a latter-day Renaissance man with a wide range of interdisciplinary interests that were too independent of the straitjacket of recognised academic disciplines and curricula, Geddes stood out against the modernists who wanted to destroy the man-made world as it then was – at least, its cities – and start all over again.

Geddes sought – and largely failed – to dispel fears of cities and mass urbanisation, and to release creative rather than destructive energies towards solving the problems of the modern city. Indeed, almost a lone voice in the wilderness, he warned against and predicted the consequences of the wholesale demolition of city centres and their urban housing, the dispersal of well-established communities and the accompanying destruction of their socio-cultural heritage, including the supportive, extended family, decades before the 1960s when the all too frequently disastrous effects of the mainstream of modern town planning became widely recognised on both sides of the Atlantic.

Geddes did, however, attract a small and devoted group of disciples during his lifetime, and he continues to be evoked as an inspiration to ecologists and conservationists alike. Certainly, many of his ideas are at least as relevant today as when he was propounding them a hundred years ago.

Geddes is acclaimed by some as the father of modern town planning. It would be tempting also to credit him as the father of urban conservation, but that would be to ignore the lack of any direct linkage – including in his adopted city of Edinburgh – and the submergence of his broad synoptic vision of cities – especially in his native land. Also, there is a better claimant.

Gustavo Giovannoni

Gustavo Giovannoni (1873–1947) trained as an engineer, architect, architectural historian and restorer. He spent his working life in Rome where he practised as an architect-planner and variously taught architecture and then architectural restoration.

Credited with inventing the term *urban heritage* and the concept of *living conservation* (in the sense of functional), Giovannoni insisted that an architect-planner needed to combine three attributes: scientist, artist and humanist. His prolific – but characteristically chaotic – published output focused directly or indirectly on a single debate: the interrelationship between the modern and the historic city at all levels: from the theoretical to the detail of practice.

The conclusion that Giovannoni explored was based on the principle of mutually supportive and harmonious coexistence. He argued that neither assimilation nor amalgamation was appropriate, and that the correct response to the question was to understand and work with the respective, complementary qualities and opportunities of each.

Giovannoni characterised the historic city by its compactness; the pedestrian pace and rhythm of life; the small scale of its urban grain, its buildings and its public spaces; the close proximity of its many different activities; and its contextual homogeneity (Figure 2.11). He characterised the modern city by its possibilities of limitless expansion; its faster pace and dynamism related to non-pedestrian forms of movement; the larger scale of its urban layout, buildings and spaces; its lack of contextuality and hence its freedom from design constraint.

He saw no need and every reason not to superimpose new quarters on top of old ones, and was a relentless opponent of Le Corbusier's ideas for the destruction and rebuilding of historic cities, which he regarded as simplistic and outdated. Giovannoni foresaw the historic area of a city as a vibrant, closely interlinked component of its new, enlarged form, performing an

Figure 2.11 Siena, Italy, typifies the historic city as defined by Gustavo Giovannoni with its human scale and contextual homogeneity. Giovannoni foresaw the historic areas of cities performing an important, distinctive and complementary role within the life of expanding modern cities.

essential and distinctive socio-economic role in the daily life of its citizens. He had no time for the notion that a historic area should be set apart from contemporary life and embalmed for historical, aesthetic and tourist objectives, and opposed the idea that it should be the subject of museological protection.

Considering the historic fabric of an old city, Giovannoni made no distinction between the recognised monuments and the modest vernacular architecture that interlinked them. He saw the two as inseparable parts of a whole, neither being complete without the other – whether contextually or functionally. He was fully supportive of controlled interventions aimed at adapting and modernising the buildings of historic areas to contemporary life, but envisaged such areas as the focus for small-scale mixed-use activities and was opposed to the introduction of incompatible large-scale activities, even if they were prestigious ones, as they would inevitably be destructive.

Giovannoni perceived it as essential that architects should have an integrated training, one from which they would be as conversant with traditional building techniques as with technical innovations, masters of historical architecture and urban form, attuned to contemporary social needs, and respectful towards context when designing new structures (Figure 2.12).

Giovannoni was a seminal figure in determining the scientific basis for the restoration of historic buildings, the adaptation of historic cities and areas, and the course of urban planning in Italy in the years between and following the First and Second World Wars. Additionally, he was a member

Figure 2.12 Urbino, Italy. Gustavo Giovannoni's ideas for the evolution of historic cities – with the sensitive extension of their historical layers employing contemporary design and modern materials – were notably taken up by the architect Giancarlo De Carlo in Urbino.

Bloomsbury: *Woburn Square* (*built 1829; Robert Sim, builder*).

Bloomsbury: *Woburn Walk, a shopping passage* (*built 1822–25; Thomas Cubitt, builder*).

Figures 2.13 and 2.14 Bloomsbury, London, England, was built between the late-1660s and the early-1830s as a largely self-sufficient suburb. Socially mixed, it included its own markets, and principal and minor shopping streets. Parts of the area were redeveloped by the University of London from the 1960s onwards. These photographs were taken in 1969.

of the conference that formulated the 1931 *Athens Charter* and represented his position during its proceedings.

Giovannoni first set out his main thesis in 1913, which suggests a remarkable originality and precocity in his ideas. Although it is not clear what direct or indirect connection existed between him and Geddes, they shared a similar evolutionary approach to cities based on analogies with biology and organic growth, botany, selective pruning and controlled regrowth.

In the United Kingdom, outside London and the major planned developments on the great private estates to the north and west of the city (Figures 2.13 and 2.14), the principal example of complementary urban development is the historical one of the late-eighteenth- to mid-nineteenth-century development of the New Town of Edinburgh alongside the Old Town. Elsewhere and since, superimposition has been the norm.

Urban conservation: mainstream beginnings in the United Kingdom

The 1960s were formative years for the future of historic cities in Britain and attempts at reconciling the emerging agenda of urban conservation with the mainstream of modern town planning. It was a decade that set the scene for debates that remain with us today, and although it produced some useful ideas and examples, in practice it aggravated more problems than it solved.

The Buchanan Report

Traffic in Towns – known officially as the *Buchanan Report* – was first published in 1963. At the time, it was interpreted in as many different ways as it had readers: from environmentalists who perceived it as a clear warning, but one that needed to be read as much between the lines as along them; to inheritors of the modernist mantle who saw it as the justification for major ongoing programmes of urban reconstruction.

The ambivalence of the report is neatly summed up by comparing the introductions to the two standard editions. Referring to the motor car in the first hardback edition, the report of the Steering Group reads: 'we are nourishing at immense cost a monster of great potential destructiveness'. By contrast, the preface to the shortened, popular, paperback edition, published a year later and written by the chairman of the same Steering Group, bolsters the need 'to replan, reshape, and rebuild our cities' and states that the importance of the *Buchanan Report* is that it shows how this can be done. This preface goes on to eulogise the virtue of rebuilding Britain's 'depressing' nineteenth-century cities and concludes: 'What the Victorians built, we can surely rebuild'.

The core question that the *Buchanan Report* sought to address was how to reconcile mass car ownership and efficient accessibility for vehicles with good environments for people – including as pedestrians – whether through the adaptation or reconstruction of cities.

The report explored the idea of *environmental areas*, within which traffic would be subordinated to the environment and through movement banned. Specifically in relation to existing cities, the report also introduced the concept of *environmental capacity*, which it defined as 'the volume and character of traffic permissible in [a] street consistent with the maintenance of good environmental conditions'. Establishing the yardstick for any given street (or, by extension, group of streets) would then determine the degree to which traffic should be restricted or eliminated altogether (Figure 2.15).

As a generic concept, *environmental capacity* has far broader relevance for historic cities than the single issue of traffic in streets.

The *Buchanan Report* included indicative studies for a small town, a large town, a metropolitan area and a historic one. Norwich was taken as the example for the historic city. The indicative solution, which focused on protecting the historic area within the medieval walls only, denying cross-movement within it by vehicles and enhancing the conditions for walking, assumed the construction of a new inner ring road, from which only restricted vehicular access to the centre would be allowed through a number of strategically placed 'gates'. Relationships to the city outside the ring were not considered.

From the 1940s onwards, major new road proposals threatened to impact seriously on historic cities up and down the United Kingdom, and by the

Figure 2.15 Beverley, England: the market square. The *Buchanan Report* recognised that good environmental conditions for people would require the restriction or elimination of traffic in existing cities, and introduced the concept of *environmental capacity* as the yardstick against which this could be measured.

1960s commitments to many such schemes had already been made: in local plans and financially. Inner ring roads, for example, now encircle the historic centres of many cities – including Norwich – isolating them visually, physically, and in most cases functionally from the rest of their cities. Such roads pre-suppose a loss of proximity between the constituent activities of citizens' everyday lives, and necessitate that a substantial proportion of the daily movements by people and goods within cities are made by motor vehicle.

The four historic city studies

The four Studies in Conservation – known at the time as the four historic town reports – that were commissioned by government in 1966 and published in 1968, addressed the English cities of Bath, Chester, Chichester and York.

The studies focused on the areas within each city's historic walls – Roman, Saxon or Medieval – with the exception of Bath, where some adjoining areas of later date were also included. As is typical throughout the United Kingdom – and continuously supported by national policy and local plans – the oldest and environmentally most sensitive parts of the four cities are also their modern commercial centres.

The purpose of these reports was stated in bald terms: 'to discover how to reconcile our old towns with the twentieth century without actually knocking them down'. Historic cities were recognised as a great cultural asset and also – but only in the context of tourism – as an economic one. The studies were conceived as pilots, and the objective was twofold: to seek solutions to specific problems and to guide parameters that could be applied nationally.

The timing of the four studies coincided with the passing in 1967 of the Civic Amenities Act, which empowered local authorities to designate conservation areas. In a very real sense, therefore, the studies were exploring what the United Kingdom notion of conservation area meant, and how to reconcile the exhortation in the Act to preserve and enhance with all of the access needs and development pressures of a modern commercial city centre.

By the mid-1960s, all four cities shared a number of key problems in their central core areas: poor maintenance and partial decay of the historic building stock; widespread disuse and accelerating dilapidation of the formerly residential upper floors of commercial properties; all-pervasive traffic – with consequent congestion and conflict between vehicles, buildings, and people; low architectural design standards in the construction of new buildings; and low levels of investment in the urban environment as a whole.

Bath

Colin Buchanan and Partners, transport planners, were already employed as consultants by the city council, and their planning and transportation study was published in 1965. The report of that study defined the environmental areas within which traffic was to be kept to a minimum, a result that the report insisted could only be achieved through the construction (or adaptation) of a hierarchy of primary, district and local distributor roads across, under and adjacent to the city centre. It was a scheme that included the construction – at very considerable financial cost – of an east–west tunnel to carry the London to Bristol A4 trunk road under the lower slopes of Lansdown Hill immediately above the north side of Queen Square, and an east–west cut-and-cover local distributor road immediately below its south side (see Figure 1.4 at page 5).

For the central area, this 1965 study focused on three main objectives: efficient access for vehicles; improved conditions for pedestrians; and the identification of redevelopment opportunities. The report admitted the merit of encouraging residential uses in the city centre on grounds of proximity, but then countered this by allocating redevelopment land for other uses – especially for the several car parks that were deemed essential in order to support the city centre commercial uses. The obligation to travel took precedence over the option of proximity.

Buchanan's 1968 conservation study sought to coincide the previous conclusions with its understanding of the expectations of statutory listing and of the Civic Amenities Act. Thus, whereas this study acknowledged the significance of the city's historical associations – notably with society in the eighteenth century – it represented the architectural importance of the city today primarily as an 'urban stage set', characterised first and foremost by the top grade Georgian facades and the urban and landscape spaces they enclose (Figure 2.16). As such, the importance of the architectural fabric of the organically developed Roman through medieval to eighteenth- and nineteenth-century city centre was downgraded in importance. The conservation report also had a highly selective approach to historic building interiors.

One of the most important issues that the Bath conservation study examined was the question of disused upper floors – a problem that has continued to affect historic city centres across the length and breadth of England since at least the end of the Second World War. Research in Bath confirmed that over forty per cent of the floor space above ground-floor commercial premises was completely disused, and much of the rest was only in nominal use. This was seen as representing a major opportunity to bring life back into the area.

Buchanan was concerned to demonstrate that these floors could be brought back into use for purposes closely related to the short- and long-term

Figure 2.16 Bath, England. The Study in Conservation for the city of Bath represented the architectural importance of city today primarily as an 'urban stage set'. Royal Crescent (built 1767–75; John Wood the younger, architect) is singled out as one of its most important components. Built to a semi-ellipse on plan, its continuous 200 metre long facade fronts thirty independent houses.

residential ones for which they had been built, namely, that they could be converted to student accommodation, hotels, and self-contained flats. This represented a positive response to the principle established in the earlier report without impinging on the opportunities for constructing multi-storey car parks on land adjacent to the central area. Although the sample architectural solutions illustrated in the conservation report were insensitive and expensive, the seed was sown for the potential return of traditional urbanity to Bath's central area.

Buchanan also applied the concept of *environmental capacity* that he had introduced in *Traffic in Towns* to the existing street pattern in the city centre, and put forward detailed proposals for environmental enhancement through traffic management, pedestrianisation, and hard and soft landscaping in the public domain.

In the event, the destructive impact of the major new road proposals that were introduced and endorsed in Buchanan's two reports, coupled with the separate but extensive slum clearance programme that was still in full swing in the city in the second half of the 1960s, fuelled the *Save Bath* campaign, and by the mid-1970s the road building proposals were largely abandoned, the slum clearance programme totally, and less destructive solutions were sought in order to encourage residential life back into the city centre.

Chester

The conservation study for Chester was undertaken by Donald Insall and Associates, noted conservation architects and planners. The most significant of its recommendations were focused on historic building and townscape issues.

The Chester study addressed the reuse of upper floors in great detail, and was the most successful of the four historic town reports in terms of its direct follow-up in the city and the major benefit it brought to the fabric and reuse of its historic buildings.

As the result of its study of the townscape of the city, the report introduced a new generic order of townscape values – anchor, major, minor, group and location – which substantially advanced the aesthetic and morphological approach to urban design generally, both in relation to the various townscape values that can be attached to existing buildings beyond those of architectural or historic merit and as guidance for the design of new buildings in a historic area.

By 1968 the population of Chester's city centre had, since 1946, fallen from 3,000 to 1,000, and the Insall report also put forward a number of proposals for the insertion of new housing into underused city centre sites.

Chichester

The conservation study for Chichester was prepared by the county planning officer. Of the four cities, Chichester was the smallest, the wealthiest and the one with the fewest problems of dereliction. The study was closely linked to a proposed outer bypass, inner ring road, and the planned closure of the east–west, north–south road intersection at the market cross close to the cathedral, all of which were subsequently implemented.

Most of the older buildings in the study area were originally domestic in use, but by the date of the study the population had fallen to just 170. The report promoted the conversion of upper floors back to residential use, but its proposals for the improvement to the lanes behind the principal streets for vehicle service access and the conversion of gardens into car parks limited the residential options.

York

At the time of its publication, the York conservation study was the most publicly acclaimed of the four historic town reports and the least well

Figure 2.17 York, England. The Esher report was the only one of the four historic city studies to challenge the merits of constructing major new roads in the vicinity of the historic city centre. By the mid-1970s, and following a sustained campaign at local and national level, the scheme to construct an inner ring road just outside the city's medieval walls was abandoned.

received by the then city council – not least because it did not support the council's plan to build a new inner ring road (Figure 2.17).

Prepared by Lord Esher, a former president of the Royal Institute of British Architects, the York report sought to analyse the walled city both from an architectural perspective and as a working and living environment. The report projected shopping as one of the mainstays of the economy of York, pitching the city centre as a major regional player in the retail sector, and it proposed improved access both for shoppers and visitors through the construction of a number of multi-storey car parks and widespread pedestrianisation. The report supported a considerable increase in shopping space, primarily in large-scale developments, and regarded the continuing success of the retail sector as essential to the prospects of survival of the fabric of the city centre – whilst ignoring the self-evident conflict.

By 1968, the population of the walled city had reduced from its peak of 10,000 in the Middle Ages to 3,500, and one of the study's core proposals was to recover this to a level of 6,000 through redevelopment by the private sector of industrial land and the backlands between street frontages. In addition to its focus on shopping and housing, the study proposed overall environmental enhancement, and anticipated that the combined effect of all three would be to achieve the conditions whereby the historic buildings would become self-conserving.

The Esher report's proposals were contingent upon the exclusion from the walled city centre of what it categorised as 'non-conforming uses'. The main culprit was identified as the 232,000 square metres of industrial-type floor space. This occupied an overall total of 10 hectares of land across the

Figure 2.18 York, England. The Esher report categorised industrial-type uses, including small workshops, as 'non-conforming', with the result that much of the vitality of the walled city has been eliminated and its service industries lost.

study area, and the report anticipated that its removal would cause 'comparatively little dislocation'.

One of the many unsatisfactory outcomes of the Esher report was that traditional buildings crafts of the type essential to maintain York's historic fabric were included in the all-embracing sanitisation that resulted from this recommendation (Figure 2.18).

The character of towns

Gordon Cullen's treatise *Townscape*, coupled with the townscape references in Donald Insall's *Chester: A Study in Conservation*, was taken up notably by Roy Worskett.

Published in 1969, *The Character of Towns: An Approach to Conservation* focused on the archaeological, architectural and visual qualities of the physical environment. Urban conservation was seen in terms of identifying and making best use of a town's features from all ages and establishing visual disciplines and design codes as the basis for managing change.

The principal impact of this 1960s' emphasis on townscape was to consolidate the view that a historic area or city's character and identity, its sense of place and distinctiveness, could and should be distilled and encapsulated simply in terms of its morphology and aesthetics, its buildings and spaces. The particularity of human activities – the noises and smells, the bustle and socio-economic interchanges – of any one urban environment when related to another were not incorporated into this simplistic equation (Figure 2.19).

In the modernist planning context of the time, this limited approach coincides with the Bath conservation report's representation of the importance of the historic architecture of that city essentially as theatrical scenery – but without the performers – and the York report's insistence that the numerous workshops, factories and warehouses be expunged from the walled city as 'non-conforming'.

Writing a few years later – by which time he had been appointed the city architect and planning officer at Bath – Worskett emphasised the importance of townscape appraisals as a management tool for assessing the capacity for visual change in a historic city. He also confirmed that he was less concerned about the uses to which buildings were put: echoing Buchanan, townscape values were more important than functional continuity, or the relationship between proximity and the choice not to travel.

One must assume that no irony was intended, but Lord Esher's conservation report for York begins with the following quotation (attributed to the Dutch architect Aldo van Eyck):

> This much is certain: the town has no room for the citizen – no meaning at all – unless he is gathered into its meaning. As for architecture; it need do no more than assist man's homecoming.

Urban Planning Context 45

Figure 2.19 Venice, Italy. The morphological and aesthetic approach to urban character does not take into account the particularity of human activities, noises and smells.

Chapter 2: digest

The mainstream of modern town planning is rooted in an anti-urban tradition that owes its origins to the harsh conditions of nineteenth-century cities at the time of the first Industrial Revolution – especially in matters of housing, coal-induced pollution and health.

This experience inspired two parallel modernising visions, of which the notable proponents were Sir Ebenezer Howard in England and Le Corbusier in France. Although they articulated their visions differently, they were based on the common principle of the separation of daily functions – residence, leisure and work – into discrete areas (or zones) connected by mechanisms of transport. The modernising vision is a top-down one that seeks to define a uniform solution to the multiplicity of urban situations.

An opposing pair of visions was expounded by Sir Patrick Geddes in Scotland and Gustavo Giovannoni in Italy, visions that recognised the distinctive civic merits of the pre-industrial city and sought to promote the integration of historic areas into modern-day life through a complementary, evolutionary process that is essentially bottom-up: its starting point is the individuality of each and every urban society and form. As an architect and town planner, Giovannoni was more focused in his approach, and his ideas were insinuated into urban planning and conservation policy in his native land.

In the United Kingdom, the legacy of separate land-use zoning has led to the concentration of a limited number of medium- to large-scale activities – primarily offices, shops and certain leisure uses – in city centres, thereby focusing the most volatile commercial uses and development pressures into the most sensitive historic areas.

Within the mainstream vision, and in consort with post-1945 programmes of city centre redevelopment that in several British cities are now well into their third generation, the *Buchanan Report* and the four historic city Studies in Conservation sought to reconcile major opposing forces without questioning the underlying thesis. This thesis pre-supposes that the qualities of historic cities are recognised primarily in morphological and aesthetic terms rather than for their traditional socio-economic activity, balance and diversity – as objects that can be characterised for their physical attributes rather than places with distinct cultural identities.

Where Le Corbusier, especially, was iconoclastic towards historic cities, Geddes and Giovannoni sought to work with them, in all of their physical and human complexity. Their approach was intrinsically conservative and economical, whilst embracing constructive progress and innovation. Geddes, additionally, recognised the need for a balanced relationship between urban societies and the natural world.

Chapter 3
Sustainability: Background

Sustainability: beginnings and evolution

In the broader, ecological sense, *conservation* and *sustainability* share the same generative basis as the mainstream of modern town planning, namely, the forces unleashed by the Industrial Revolution and the associated serious environmental consequences of the loss of equilibrium between the human and natural worlds.

The starting points for concern are numerous. They include modern warfare; population growth; deforestation and desertification; loss of habitat, animal species and biodiversity; drought and famine; diminishing reserves of natural resources; toxic wastes and air pollution; industrial accidents; acid rain and ozone depletion; global warming and climate change; and health and global equity.

Reactions vary from the apocalyptic, through the fatalistic and tunnel-vision, to the positive, holistic, problem-solving approach.

The apocalyptic prognosis is not new. Thomas Malthus (1766–1834), writing at a time when the world population was barely one billion, predicted that population growth would eventually outstrip the means of subsistence, resulting in catastrophe. In 1968, at a time when the world population had reached three-and-a-half billion (it is now six-and-a-half billion and rising), Paul Ehrlich reignited this argument in his book *The Population Bomb*.

Over the same two-century time frame, authors of fiction have also been painting nightmare visions: from Mary Shelley (1797–1851); through Aldous Huxley (1894–1963; grandson of Sir Patrick Geddes' mentor, Thomas Huxley, and brother of the biologist Sir Julian Huxley, the first Director-General of the United Nations Educational, Scientific and Cultural Organisation (UNESCO)); to George Orwell (1903–50; once a pupil of Aldous Huxley).

The focus of concerns has changed and accumulated over the decades since the Second World War.

Through the 1950s, leading naturalists increasingly drew public attention to ecological issues relating to habitat and to the near extinction of many

Loss of habitat, animal species and biodiversity: raising the profile

The trio of British naturalists who came into prominence in the 1950s

Sir Peter Scott (1909–89), ornithologist, conservationist, painter, broadcaster and author, founded the Severn Wildfowl Trust (now the Wildfowl and Wetlands Trust) at Slimbridge in Gloucestershire in 1948. He was one of the founders of the World Wildlife Fund (WWF) and designed its panda logo.

Gerald Durrell (1925–95), naturalist, author, broadcaster and educationalist, is best known for founding the Jersey Zoological Park in the Channel Islands in 1958, the Wildlife Trust that now bears his name, and for his many books based on his family, his animal-collecting expeditions, and his conservation efforts. Durrell sought to highlight the increasingly serious loss of biodiversity in the world through loss of natural habitat and species, and was a pioneer in the field of the captive breeding of endangered animals (Figure 3.1).

Figure 3.1 Early endeavours to alert the world to ecological threats were focused on the natural world. The World Conservation Union (or International Union for Conservation of Nature and Natural Resources, IUCN), founded in 1948 and with its headquarters in Gland, Switzerland, opened its *Red List of Threatened Species* in 1963. The *IUCN Red List* is recognised as the most authoritative gauge of the status of conservation and biodiversity in the world. The May 2006 update listed 16,118 threatened species (out of an overall global total of 40,168), of which 7,725 were animals and the balance plant forms. Gerald Durrell regarded captive breeding as a holding measure, and was involved in the return of several endangered animal species to the wild. (La Montagne des Singes, Kintzheim, France.)

Sir David Attenborough (born 1926) is one of the world's best-known broadcasters and naturalists. Attenborough made his first appearance on television with his series *Zoo Quest* in 1954. Widely considered one of the pioneers of the nature documentary, he has written and presented several major series surveying nearly every aspect of life on Earth. His television series *State of the Planet* in 2000 examined the environmental crisis that threatens the ecology of the planet and concluded with the following warning:

> The future of life on earth depends on our ability to take action. Many individuals are doing what they can, but real success can only come if there is a change in our societies and our economics and in our politics. I have been lucky in my lifetime to see some of the greatest spectacles that the natural world has to offer. Surely we have a responsibility to leave for future generations a planet that is healthy, inhabitable by all species.
>
> In May and June 2006, respectively, he broadcast the two-part environmental documentary *Are We Changing Planet Earth?* and *Can We Save Planet Earth?*

species in the wild. Over succeeding decades they represented ever more forcefully the view that other species have just as much right to inhabit the Earth and enjoy quality of life as the human race, and that we are all part of the same, mutually interdependent and supportive ecosystem. Meantime, in 1952 London witnessed the worst single air pollution incident in United Kingdom history, one that served as a major alert about the causes and effects of urban pollution and stimulated the first tentative steps to address it.

Urban pollution

The causes and effects of air pollution

The causes of air pollution can be chemical, physical (including particulate) or biological, and the connections between emissions and acid rain, ozone depletion and global warming are well documented. Even discounting serious individual industrial accidents – such as the escape of toxic gases from a chemical plant in Bhopal, India, in 1984, or the radioactive release and fallout from the nuclear power station at Chernobyl, Ukraine, in 1986 – air pollution is the substantive contributory cause of a significant number of fatalities each year (estimates suggest the figure may be as high as three million world wide, including over 300,000 in Europe) and respiratory disorders.

The Great London smog of December 1952

The smog that settled over London in December 1952 had a catalytic impact on public awareness of the relationship between air pollution and health, and stimulated scientific recommendation and government regulation.

Smogs were commonplace in major industrial cities in Britain from the first half of the nineteenth century well into the second half of the twentieth. They occurred especially in periods of calm, damp, autumnal or wintry weather, when the sulphur-laden smoke particles from industrial, commercial and domestic consumers of coal combined with fog and the glow from street lighting to form a thick yellow-black, all-pervasive 'smog'. The pollution levels of a smog are highest at ground level.

London smogs were made famous by Sir Arthur Conan Doyle (1859–1930), author and creator of the fictional detective Sherlock Holmes, and many visitors came to see the capital in the decades before the First World War – at the height of the British Empire – to experience these smoke-laden fogs.

The weather conditions of early December 1952 were particularly favourable, and an especially dense smog blanketed the entire city and its suburbs for several, windless days. The levels of smoke in the atmosphere increased threefold; of sulphur dioxide, sevenfold. Research has indicated that there were 13,500 more deaths than normal for the period December 1952 to March 1953, of which 4,500 occurred during the week when the smog was at its peak; research also indicates that the mortality rate did not return to normal levels for several more months.

As a child in my first term at primary school, I recall that visibility was quite literally nil, that street lights appeared merely as a faint glow from an imprecise direction and at an indeterminate height, and that the only way home was by a combination of sense of direction to cross streets, and by touch of garden walls, fences and gates to feel one's way step-by-step along the inside edge of pavements.

The British government passed the first of its Clean Air Acts in 1956, introducing smokeless zones. Over time, across the United Kingdom, coal was phased out in industry, the office, and the home (Figure 3.2). Coal remains a major source of energy for industrial production and a continuing cause of air pollution in countries such as China.

Figure 3.2 The familiar name for the city of Edinburgh, Scotland, is *Auld Reekie* – in recognition of the coal-induced air pollution that characterised its skies until the Clean Air Acts came into force. The uncleaned monument to the writer Sir Walter Scott (1771–1832), erected to his memory in Princes Street in the 1840s, bears gaunt witness to the age of coal. Today, cyclists in the city wear face masks to protect themselves from the carbon emissions from motor vehicles.

Urban pollution today

Photochemical smogs – as they have been described since the 1950s – affect most major urban centres and their surrounding areas in the world: from Asia across the Americas to Europe; affecting cities as diverse as Beijing, Los Angeles, Mexico City, Athens, and – albeit from a different source than the great smog of half a century ago – London. Photochemical smogs are worse during periods of calm and hot summer weather. Modern industry is a contributory factor, but the greatest single source of concentrated, harmful emissions in urban areas today is motorised traffic.

Urban pollution affects hundreds of millions of people in cities across the continents. The 1952 London smog, a major incident at the time, pales into insignificance compared to the magnitude of the problem today.

The 1960s were principally significant for events emanating from North America. It was the decade of 'flower power', 'make love not war', and other expressions of the ideology of non-violence; of civil rights' movements; and of increasing environmental awareness.

Silent Spring, published in 1962, first drew attention to the environment's limited capacity to absorb pollutants. Friends of the Earth was founded in the United States in 1969, became an international network in 1971, and acts as a forceful advocate of environmental responsibility, cultural and ethnic diversity, and civil empowerment.

Greenpeace was founded in Canada in 1971, now has its international headquarters in Amsterdam, and is described in its official mission statement as

> an independent, campaigning organisation which uses non-violent, creative confrontation to expose global environmental problems, and to force solutions for a green and peaceful future. Greenpeace's goal is to ensure the ability of the earth to nurture life in all its diversity.

Greenpeace acquired and named its first ship the *Rainbow Warrior* in 1978.

The 1970s saw an increasing level of awareness of global environmental issues and concerns about energy supplies.

The Club of Rome, a global think tank whose members include scientists, economists, businessmen, administrators and statesmen, aims to inspire individuals to take responsibility for societal improvement in the world. Founded in 1968, the first of its series of commissioned reports was published in 1972. *The Limits of Growth* modelled the dynamic interaction between industrial production, population, environmental damage, food consumption, and the usage of finite natural resources, and predicted that economic growth could not continue indefinitely. The 1973 oil crisis, induced by the Organization of the Petroleum Exporting Countries (OPEC), added credibility to this argument at the time.

Critics have sought to discredit the Club's repeated warnings as too pessimistic. Economic growth is, of course, the cornerstone of business and

Key United Nations initiatives in the field of sustainability

1972 United Nations Conference on the Human Environment

The Conference held in Stockholm, Sweden, in 1972 brought together industrialised and developing nations and marked the formal acceptance by the international community that development and the environment are inextricably linked. It prompted a growing body of scientific research into critical environmental issues, and led to the establishment of the United Nations Environment Programme, based in Nairobi, Kenya, the same year.

1987 Montreal Protocol on Substances that Deplete the Ozone Layer

The *Montreal Protocol* is an international treaty designed to protect the ozone layer by phasing out the use and production of directly harmful substances (Figure 3.3). Since the protocol entered into force in 1989 the atmospheric concentrations of the most important chlorofluorocarbons (CFCs) have either levelled off or decreased.

The *Montreal Protocol* was described in 2000 by Kofi Annan, Secretary-General of the United Nations, as 'perhaps the most successful environmental agreement to date'.

Figure 3.3 The Ozone Layer. Ozone concentrations are greatest between 15 and 40 kilometres above the earth. Most commonly linked to the risk of skin cancer in humans, the ozone layer protects all life forms in the biosphere by absorbing biologically harmful ultra violet radiation emissions from the sun. Civil aircraft normally fly at a maximum height of around ten kilometres, below the lower limit of the stratosphere.

1987 Our Common Future (also known as the Brundtland Report)

The World Commission on Environment and Development was established in 1983, and Gro Harlem Brundtland, thrice Prime Minister of Norway in the period 1981–96 and Director-General of the World Health Organisation from 1998 to 2003, was appointed to lead it. *Our Common Future*, the Commission's report, popularised the term *sustainable development*. The report highlighted the conclusion that current patterns of resource consumption and environmental degradation cannot continue and that economic development must adapt to the planet's ecological limits.

The *Brundtland Report* set out three fundamental components to sustainable development: environmental protection, economic growth and social equity (Figure 3.4). The report was concerned with securing global equity – enabling developing nations to meet their basic needs of employment, energy, water and sanitation whilst encouraging their economic growth – and recognised that achieving equity and sustainable growth would require technological and social change.

Figure 3.4 The *Brundtland Report* highlights the need for economic development in the Third World to overcome poverty, high birth rates and environmental degradation. (Eritrea, Horn of Africa.)

The conclusions of the *Brundtland Report* were echoed subsequently, and less ambiguously, in statements attributed to Pope John Paul II. First: 'Modern Society will find no solution to the ecological problem unless it takes a serious look at its life-styles'. Second: 'From now on it is only through a conscious choice and through a deliberate policy that humanity can survive'.

1992 United Nations Conference on Environment and Development (also known as the Earth Summit)

The conference held in Rio de Janeiro, Brazil, was attended by delegates from some 180 countries, including over a hundred heads of state. At the time, it constituted the largest environmental conference ever held.

The *Earth Summit* is notable for the five main agreements that came out of it, agreements that aimed to cover every aspect of sustainability: the Convention on Biological Diversity; the Framework Convention on Climate Change; Forest Principles; the Rio Declaration on Environment and Development; and Agenda 21 – the blueprint for sustainable development in the twenty-first century.

The Rio Declaration stated that 'humans . . . are entitled to a healthy and productive life in harmony with nature', stressed the importance of eradicating poverty, of international cooperation, and the need to 'conserve, protect and restore the health and integrity of the Earth's ecosystem'. The Declaration also expressed the need to preserve cultural identity within the development process.

Following the 1992 *Earth Summit*, the United Nations established the International Commission on Sustainable Development to monitor the implementation of Agenda 21 and to support the achievement of its objectives through the involvement of civil society at local level.

The Rio *Earth Summit* generated a sense of optimism that adherence to the principles of sustainability was gathering substantive momentum at a global scale.

1997 Kyoto Protocol to the United Nations framework convention on climate change

The *Kyoto Protocol* is an international treaty on climate change. Arising directly out of the 1992 *Earth Summit* held in Rio de Janeiro, it is intended to represent a commitment by industrialised nations to the reduction of greenhouse gas emissions – especially carbon dioxide – with the view to stabilising their effect on global warming. The *Kyoto Protocol* entered into force in 2005; to date, it has a long way to go before it achieves the same level of active support as the 1987 *Montreal Protocol*.

2002 World Summit on Sustainable Development (also known as the Earth Summit 2002)

The conference held in Johannesburg, South Africa, was intended as a statement of ongoing, enhanced commitment to the programme of initiatives dating back to the 1972 Stockholm Conference – especially, the Rio *Earth Summit*.

The Johannesburg Conference was, however, dogged by controversy and a sense of impotence. Progress in implementing sustainable development had been disappointing since the euphoria of the 1992 *Earth Summit*, with poverty deepening and environmental degradation worsening in many parts of the world – not least on the African continent where the conference was held.

economic politics around the world, whether in those regions of the northern hemisphere that enjoy continuity of it or those in the southern hemisphere that seek to emulate it.

The year 1972 also witnessed the first of the major series of international conferences, protocols and other initiatives emanating from the United Nations, initiatives that have triggered parallel waves of regulation and governmental policy instruments across Europe and placed sustainability as a core agenda of our times.

The changing focus and accumulating priorities may be summarised by characterising the 1980s in general terms as the decade of energy audits, the 1990s of environmental assessments (including Environmental Impact Assessments), and the 2000s as the decade of sustainability plans (including Local Agenda 21s) – in which the concepts of finite resources, life cycle, biodiversity, liveability, health and safety, and social equity have increasingly come to the fore.

Since the 1950s, *conservation* – in the broader, ecological sense – and *sustainability* have grown from an initial emphasis on the natural world and the legacy of the first Industrial Revolution to keynote international agendas and national programmes aimed at combining economic development with social and environmental responsibility at the global through to local scale (Figure 3.5).

Awareness levels have been raised at the international diplomatic level first and foremost by the United Nations, and at the grassroots activist level

Figure 3.5 Since the 1950s, environmental concerns have shifted from an initial emphasis on conservation in the natural world to embrace the over-arching concept of sustainability (South Island, New Zealand).

by the 'warriors' of organisations such as Greenpeace. Only to a more exclusive extent has *conservation* in the sense of architectural and urban conservation achieved international diplomatic support (primarily through UNESCO, its filials and offshoots), and it lacks an equivalent network of 'warriors'.

Critics will contest the detail and sceptics will question the timing, but the threats of irreversible damage to the planet appear clear, and evidence such as the melting of the polar ice caps irrefutable.

This all points to the need for a 'sustainability revolution'. Whether this is a peaceful or a violent one will depend on whether we heed today's many warning signs and act upon them more vigorously, or continue to live in a fool's paradise and await the onset of disasters.

The language of sustainability

Sustainable development, sustainability and sustainable

The 1987 *Brundtland Report* is a global agenda for change that seeks to place environmental issues to the fore. It addresses population growth and environmental degradation; health and education; adequacy of food supplies; conservation of species and ecosystems; sustainable industrial development and sources of energy; and strategies for the developing world.

The sound-bite that the *Brundtland Report* is best recognised by is its popularisation of the term *sustainable development* and its simplistic definition of it: 'Sustainable development is development that meets the needs of the present without compromising the ability of future generations to meet their own needs'. It is a definition that emphasises development and is underscored by a commitment to social equity between and within generations.

This definition has spawned well over 200 others that variously paraphrase, clarify or apply it to particular situations. It is, for example, interpreted to mean development that utilises renewable resources at rates less than their natural rate of regeneration or, contrastingly, as development that optimises the efficiency with which non-renewable resources are exploited.

Since it was published, the simplistic Brundtland definition has increasingly attracted criticism from a number of quarters. Some remark that it is focused towards human needs and ignores the needs of the natural world. Others object to the emphasis on development, suggesting that *sustainability* and *development* are as likely to be in tension as they are in harmony, and that the term *sustainable development* should be reserved for specific rather than general activities.

The most useful distinction that may be drawn between *sustainability* and *sustainable development* is to consider the first as a positive and continuous over-arching process for all human activity, one that binds the well-being of people and the ecosystem into a mutually supportive whole, and the second as a goal for specific situations. Thus, *sustainable development* is a potential component of *sustainability*. The commonality is that both concepts embrace, long term and holistically, environmental, social and economic issues, of which environmental protection must hold primacy as it underpins existence, or, more plainly, survival.

Multiple usages of *sustainable development* and *sustainability* abound, many of which do not accord with the distinction noted above. Development in a northern hemisphere urban sense is often predicated primarily for financial reasons. To the town centre investor, for example, whose focus of attention is commercial property, *sustainability* is employed to mean achieving a level of economic growth that is attractive to continuous capital investment into existing buildings, new development or other significant change-driven and money-orientated activity.

On the premiss, therefore, that development may be forced for economic reasons that are unrelated to social and environmental ones, and may indeed be in conflict with them, the suggestion has been floated that *sustainable evolution* is a more relevant concept to work towards than *sustainable development*. Such a suggestion would enable the connection that necessarily exists in people's minds between development and economic growth to be loosened, and substituted by the more equitable objective of human development: improvements in health and education rather than material consumption; emphasising quality of life rather than more affluent lifestyles. Evolution, as we have seen, was the central theme of Sir Patrick Geddes' work with cities and citizens.

The adjective *sustainable*, employed independently of *sustainable development*, is now so overused that some people are apt to groan just at the sight or sound of it. Historians may state that architectural conservation is by its very nature *sustainable* because it sustains architectural styles – but quite what this has to do with meeting present and future human needs, let alone with the biosphere, is unclear. Likewise, the official United Kingdom government's short definition of a sustainable community reads: 'Sustainable communities are places where people want to live and work, now and in the future'. In this definition, it would be more meaningful to delete the word *sustainable* and substitute *happy*.

In short, *sustainable* is applied freely to any idea, policy or product in order to justify it, enhance its appeal, or market it in a world where political correctness increasingly demands that all and everything is seen to be environmentally friendly – or, using today's catchword, *sustainable*.

If this all sounds confusing, so it is. More relevant than the architectural historian's response is that of one of the United Kingdom's foremost conservation architects, Sir Bernard Feilden, namely, that *sustainability* is about prolonging the useful life of a building in order to contribute to a saving of energy, money and materials. This establishes a clear relationship with the finite resources of the natural world, and successfully embraces the three components of *sustainability*: environment, society (functionality within the community) and economy. And more relevant than the short definition of sustainable community quoted above is the long one:

> Sustainable communities are places where people want to live and work, now and in the future. They meet the diverse needs of existing and future residents, are sensitive to their environment, and contribute to the quality of life. They are safe and inclusive, well planned, built and run, and offer equality of opportunity and good services for all.

This may not – in either the ecological or economic senses – be a fully inclusive definition, but at least it bears more relationship to the tripartite definition of *sustainability* than the shorter version.

As with architectural conservation, clarity and unanimity in the meaning and use of words is absent.

Relevance to historic cities

Assuming we accept that there is a global challenge to address, and that we have the responsibility and opportunity to respond to it at different levels, cities from the past – our historic cities – have a potentially significant role to play. It is not therefore a question of *why*, but *what*, *how* and *when*.

Much of the global debate about sustainability has little relevance to historic cities if they are only thought of in terms of architectural and historic interest: of ensembles of historic monuments large or small; the history of styles and techniques of construction; and the aesthetics of form, space and townscape.

If, on the other hand, historic cities are considered in terms of their functionality within communities, the natural resources of materials and energy that have gone into their construction, and the financial means that have been invested in them often over several generations, then the relationship to the three core issues of sustainability becomes more evident.

From the mainstream of modern town planning we have inherited an approach to historic cities whose starting point is confrontational: seeking to rebuild or at least very substantially remould them to a particular set of preconceived notions that are time dated. This approach may have been modified over time, but there remains a presumption in, for example, the United Kingdom's planning system in favour of development, and the majority of that development is focused in cities.

Others – notably Geddes and Giovannoni – have presented consonant approaches: ones that are directed at understanding how individual cities work within their communities, with the view to devising and implementing tailor-made strategies to suit their particular socio-economic and environmental circumstances and to taking best advantage of their inherited investments in place and people. Of course, this is a much more intellectually demanding approach than simply devising blanket solutions that are then imposed universally.

In the matter of *what*, the issues begin to stack up. They include continuity of socio-economic and environmental functionality; continuity of use of the material resources and products in infrastructure and buildings that have already been extracted and manufactured; avoidance of unnecessary use of finite reserves of fossil fuels in the transportation of goods and people; and avoidance of all related waste and pollution. Embraced within these is respect for, and continuity of, cultural identity and diversity.

Urban conservation: strategic beginnings at the metropolitan scale

It was Sir Peter Hall, doyen of planning authors, who wrote: 'The most fundamental problem of the modern metropolis is the continued increase in employment right at the centre'.

London, England

The South East Study: 1961–1981 was published by the government in 1964. It sought to address a series of issues that had become apparent since the mid-1950s, foremost among which were a projected three-and-a-half million increase in the population of the region during the period of the study; an accelerating increase in the number of mostly office-based service jobs and their concentration in the heart of the capital – in the City and the West End; a shortage of building land for housing close to the capital; a

Figure 3.6 The skyline of the City of London epitomises a monocentric metropolitan city. The City ceased to be a residential neighbourhood from the mid-nineteenth century onwards; it became the focus for office developments, deserted at night and at the weekends. Since the 1950s, height restrictions on the skyline have been eased, and many historic landmarks are now visible only at close range.

net outflow of population from London to the suburbs and beyond; and corresponding projected increases in commuter travel – which the study recognised were wasteful in time and economically inefficient for the transport infrastructure.

As a complement to the first generation of New Towns that were already being developed in the region, the study proposed a series of counter-magnets to the capital in the form of three new cities and the substantial expansion of six others, all of which were located a considerable distance from London itself. The intention was that these would balance the single focus of the centre, attracting employment and population away from it.

In the event, the major components of this plan were dismantled one by one, and commuters to London now travel daily from distances greater than the farthest of the projected counter-magnets.

London may be considered an essentially monocentric metropolitan city, and the skyline of its central areas reflects this (Figure 3.6).

Paris, France

In 1925 the architect-planner Le Corbusier presented his *Plan Voisin* for the rebuilding of Paris, a plan that included a proposal to demolish the entire quarter of the Marais (see also page 28). This plan evoked early reaction in the capital, and from the second half of the 1920s onwards debate intensified about how to protect this important historic area from redevelopment.

As a first step, all twenty *arrondissements* (districts) of the city of Paris were protected under the 1930 Sites Law. This law established a protective list of urban areas and landscapes, principally for preventative, development control reasons. Thus, since the 1930s – with some specific, isolated breaches – the urban landscape of Paris has been subject to strict skyline control.

Until the 1950s, and for at least two and a half centuries since the reign of Louis XIV (1643–1715), France was politically, economically and culturally the most centralised state in Europe. Progressively and continuously, this centralisation has been addressed at a national level. Much economic and political power has been passed to the provinces, and the promotion of a wide spectrum of cultural activities, events and venues has played a major part in this process. By the early-1960s, major centres of expansion were identified in all compass directions from Paris, including the provincial cities of Lille, Lyon, Strasbourg and Nantes – respectively north, south, east and west.

Decentralisation has also played a major part in the strategic planning of the Paris region. Five new regional centres of population and employment were also identified, including Cergy-Pontoise to the north-west and Evry to the south-east, and have been substantially successful in siphoning development pressures away from the city centre.

Additionally, and of greatest significance for the protection of central Paris from the pressures of commercial redevelopment and saturation with office-based employment, was the establishment in 1958 of the new business

Sustainability: Background 61

Figure 3.7 Paris, France. The business and administrative centre of La Défense, established in the late-1950s outside the *boulevard périphérique*, has siphoned pressures for commercial redevelopment away from the city centre and created favourable conditions for the conservation of historic areas such as the Marais.

Figure 3.8 Paris, France. The focal point of La Défense is La Grande Arche (completed 1989; Johan Otto von Spreckelsen, architect). Aligned on the axis of the Champs Elysées, it is more than twice the height of the Arc de Triomphe and four times that of the Arc du Carrousel. The location of major commercial development away from city centres enables freedom of design and scale in new buildings without impinging on the integrity of their historic areas.

and administrative centre of La Défense, some eight kilometres west of the Louvre on the axis of the Champs Elysées, outside the *boulevard périphérique* and beyond the Bois de Boulogne.

By 1990, La Défense had become Europe's largest purpose-built business centre, housing government ministries and some 650 company headquarters. Today it hosts over 1,500 French and international companies on its 170 hectare site, 150,000 employees, and 20,000 residents – with many thousands more nearby (Figure 3.7). Also, major improvements were carried out to the public transport infrastructure across the whole of the Paris region, including the construction of an additional network of deep-level higher speed metro lines – one of which connects La Défense to the Louvre in less than ten minutes.

The combined effect of these various initiatives has been to distribute the pressures for development; to balance the movements of people to and from different parts of the city and its region – thereby increasing the efficient use of the transport infrastructure; to create favourable environmental and economic conditions for the protection and conservation of the historic areas of the city; and to provide positive outlets for major new developments that do not conflict with the historic core and its buildings (Figure 3.8).

Similar strategies at different scales have been applied throughout France, all to the effect of channelling major pressures for commercial development – especially for offices and retail – away from historic city centres.

Figure 3.9 The Paris skyline epitomises a polycentric metropolitan city. Height restrictions have been in force since the 1930s; with limited exceptions they have been vigorously enforced. All of the city's historic landmarks are visible across the city; La Défense features only at a distance, on the western horizon.

The skyline of Paris reflects the fact that since the Second World War it has developed as a polycentric city, its old and new quarters in the kind of harmonious juxtaposition that Gustavo Giovannoni sought to promote (Figure 3.9).

Chapter 3: digest

The concept of sustainability has grown out of a set of initially loosely related environmental concerns that encompass issues as diverse as loss of species and habitat in the natural world; human population growth; the rate of exploitation of non-renewable natural resources; the quantity and nature of wastes and pollution; and impacts on the ozone layer and climate.

The keys to the concept are first, the understanding that all life forms on the planet are part of the same, complex, mutually interdependent ecosystem; and second, the recognition that mankind has both the responsibility and the opportunity to address the core issues in a coordinated and beneficial manner.

Since the 1970s, initiatives at the international scale have sought to respond to the issues. The *Brundtland Report* established a unified sense of direction. The *Montreal Protocol* and the Rio *Earth Summit* were notable successes. The *Kyoto Protocol* entered into force in 2005.

One of the catchphrases to come out of the sustainability agenda is: 'Think Global, Act Local'. Important recurrent themes are the importance of both biodiversity and cultural diversity.

In the context of historic cities, these signals support the view that bottom-up consonant approaches that respect the human and resource values of existing communities and places are more relevant than top-down confrontational ones that seek to impose a limited set of received ideas. This affords the opportunity for architectural conservation to play a far more central role in guiding the future of the urban environment than it has hitherto.

Just as success in the field of sustainability requires a coordinated approach at the international level, so does success in urban conservation require a strategic approach at least at the scale of the city-region. It also requires urban conservation to embrace concerns about the exploitation of natural resources, the use of energy, and the production of wastes and pollution with at least as much enthusiasm as architectural and historic interest and townscape.

Chapter 4
Conservation: International Initiatives and Directions

UN + UNESCO

1972 was the year that marked the beginning of coordinated international initiatives in the fields of both *sustainability* and *conservation*. It was the year that the United Nations (UN) headed the Conference on the Human Environment in Stockholm, and the year that its sister organisation, the United Nations Educational, Scientific and Cultural Organization (UNESCO), adopted the *World Heritage Convention* (the *Convention*) at its General Conference held in Paris.

The words *sustainability* and *conservation* did not feature pre-eminently in either of these initiatives at the time – the emphasis was on *environment* and *protection* – and the increasing complementarity of the actions of the two organisations suggests that it is unhelpful to place them into rigid compartments. Thus, to say that the UN = *sustainability* and UNESCO = *conservation* would be oversimplistic. Indeed, *sustainability* is a core agenda of both the United Nations and UNESCO, and their strategies are mutually supportive. The *Operational Guidelines for the Implementation of the World Heritage Convention* state the complementarity as follows:

> Since the adoption of the *Convention* in 1972, the international community has embraced the concept of *sustainable development*. The protection and conservation of the natural and cultural heritage are a significant contribution to sustainable development.

UNESCO

The World Heritage Convention

The Convention Concerning the Protection of the World Cultural and Natural Heritage – to give it its full title – addresses a number of parallel objectives.

First, by its very title and its 'one world' emblem – a central square representing a form created by man that opens out and is enclosed by a circle representing the natural world – the *Convention* symbolises the interdependence of cultural and natural heritage and the mutuality of their protection. The *Convention* was the first key international document to make the connection between culture and nature, and remains one of the few that do. At a national level, generally, cultural and natural heritage are sectioned and dealt with separately (Figure 4.1).

Second, the *Convention* introduced the concept of a common world heritage of 'outstanding universal value' and of the duty of the international community to cooperate to ensure its protection and transmission to future generations for the benefit of humankind as a whole. The World Heritage List and the World Heritage List in Danger are products of this objective.

Third, Article 5 of the *Convention* commits state parties (the signatory governments) to establishing effective and active measures for the protection, conservation and presentation of 'the whole of [their] national heritage, whether or not it is recognised as World Heritage'. This includes adopting:

> a general policy which aims to give the cultural and natural heritage a function in the life of the community and to integrate the protection of that heritage into comprehensive planning programmes.

Figure 4.1 France. The 1930 Sites Law established a protective list of rural and urban landscapes. Outside the more limited remit of the National Park Service in the United States of America – a model that inspired much of the debate that led to the adoption of the UNESCO *World Heritage Convention* in 1972 – the French Sites Law is the earliest nation-wide measure to embrace both cultural and natural heritage. It has no equivalent in the United Kingdom, and few parallels elsewhere in the world. (Belcastel, south of Souillac in the Dordogne.)

This commitment is reinforced by the *Recommendation Concerning the Protection, at National Level, of the Cultural and Natural Heritage*, which was adopted in parallel with the *World Heritage Convention* at the selfsame UNESCO General Conference in 1972.

This *Recommendation* emphasises the importance of providing cultural and natural heritage with an active function in the present and for the future; also, of integrating this heritage into social and economic life, and regional and national planning policy generally, to the extent that it is not regarded as a check on development, rather as a determining factor in it. Additionally, the *Recommendation* sets out the wealth that this heritage represents; the need to consider it as a homogenous whole, including its most modest components; the importance of careful and continuous maintenance for the avoidance of costly programmes to redress degradation; and the need in the historic areas of cities to ascertain the social and cultural needs of the communities situated within them alongside the town planning and technical issues of rehabilitation and conservation.

UNESCO is at some pains to emphasise the over-arching commitment under Article 5 of the 1972 *Convention* and its more expansive form under the 1972 *Recommendation*. Both, however, are over-shadowed by the focus on the World Heritage List.

There is also an unfortunate ambiguity between the official versions of the texts. The English versions of both the *Convention* and the *Recommendation* read 'protection, conservation, and presentation'. Of these, *presentation* is taken to mean *interpretation* in the sense of communicating knowledge and understanding. The equivalent in French reads *mise en valeur*, which relates far more closely to taking advantage and making best use of, including through enhancement and conservative development. The Spanish and Portuguese versions follow the French. Awareness and education are dealt with separately in both documents. In the English versions this appears to represent a duplication; in the others, it is complementary.

There is of course a significant difference between interpretation and use, between being guided as an observer of heritage and employing it in everyday life. It may be considered that much of the current emphasis in the United Kingdom on interpretation has the effect of reinforcing the limited view of heritage as 'the culture, property, and characteristics of past times' (see page 7) rather than providing it with a 'function in the life of the community' (Article 5 of the *Convention*).

The World Heritage List

The World Heritage Committee determines which sites are to be inscribed on the World Heritage List and monitors the state of conservation of those already on it.

The Committee is served by the World Heritage Centre as secretariat and by three technical advisory bodies: for cultural sites, the International Council on Monuments and Sites (ICOMOS) and the International Centre

Figure 4.2 Florence, Italy. The early entries on to the World Heritage List showed a strong bias in favour of European sites. The historic centre of Florence, symbol of the Italian Renaissance, was inscribed as a World Heritage Site in 1982.

for Conservation in Rome (ICCROM); for natural sites, the World Conservation Union (or International Union for the Conservation of Nature and Natural Resources (IUCN), the same organisation that maintains the *Red List of Threatened Species*).

In April 2006 there were 182 state parties to the *Convention*, and in July 2006 the World Heritage List comprised a total of 830 sites in 138 countries. Of the 830 sites, 644 were classified as cultural, 162 natural, and 24 mixed.

The World Heritage List is necessarily selective and subject to reflection over time. Analysis of the list of cultural heritage sites inscribed in the first two decades of the operation of the *Convention* discloses a strong bias in favour of ones that coincide with the roots of architectural conservation, namely, European, Christian, and monumental (Figure 4.2). Since 1994, an evolving global strategy has sought to achieve a more balanced list, including by geocultural distribution, by under-represented categories of sites, and between cultural and natural sites.

This process of reflection has necessarily involved the examination of a number of precepts, including the interpretation of *heritage* in different contexts.

Cultural landscape

The category of *cultural landscape* was first used in 1993 in relation to Tongariro National Park, North Island, New Zealand, a site that has strong

Figure 4.3 Ironbridge Gorge, England, pioneered the category of industrial heritage when it was inscribed as a World Heritage Site in 1986. It is a site where the interrelationship between the natural landscape, its resources, their exploitation, the evolving society, and the human reshaping of the landscape is primordial.

Figure 4.4 Bath, England: the Great Bath. Bath owes its name and its fame to its natural hot springs. They are unique in Britain and have been the formative influence throughout the city's history, from the initial settlement of the site in pre-historic times to the city's continuous development as a spa town and inland resort since the Roman period. The Roman bathing complex of hot baths, steam rooms, whirlpools and plunge pools dates from the first to the fourth centuries AD.

cultural and religious significance for the Maori people. The category is now defined in the UNESCO *Operational Guidelines* as denoting a cultural heritage site that represents the 'combined works of nature and man'. They are sites where the cultural identity and achievements of a particular society have been forged over time by interaction with their natural environment and where, generally, that natural environment has itself been modified as a result.

In the United Kingdom, a site such as Ironbridge Gorge, inscribed on the World Heritage List as an industrial heritage site in 1986, would now also be recognised as a cultural landscape (Figure 4.3).

Likewise, the city of Bath, inscribed as a World Heritage Site in 1987 primarily for its Roman archaeology and Georgian architecture, would now also be recognised as a cultural landscape: for the formative interrelationship between the natural hot springs and the city's evolution as a focus for health cures and fashionable society, and for the landscape setting as the inspiration for the originality and fluidity of its eighteenth-century town planning (Figure 4.4).

Authenticity

The definition of *authenticity* according to the elemental European concept has been subject to challenge.

Figure 4.5 Warsaw, Poland: the Market Place, focal point of the historic centre. The 1982 *Declaration of Dresden* recognised the spiritual and symbolic validity of reconstructing cultural sites following wartime destruction. The Declaration pre-supposes the existence of reliable documentation (in this case there were both photographs and drawings) and emphasises the importance of architectural development and functional continuity from the past into the future.

First, in Europe itself. The historic centre of Warsaw was razed to the ground during the Second World War. The exemplary, post-1945 reconstruction of the heart of the Polish capital was recognised by its inclusion in the World Heritage List in 1980 (Figure 4.5). As such – in a material sense – the concept of authenticity was attributed to the reconstruction, not to what had existed previously.

In a similar vein, Rila Monastery in Bulgaria was inscribed on the World Heritage List in 1983. The ICOMOS advisory report at the time did not support the nomination: it considered that the nineteenth-century reconstruction of Rila – which it describes as 'contemporary' – did not satisfy the criterion of authenticity, as very little remained of the earlier, fourteenth-century monastery. The World Heritage Committee, however, determined that the grandiose reconstruction itself represented a significant example of nineteenth-century Bulgarian Renaissance (Figure 4.6).

Second, outside Europe, in Asia for example, where the authenticity of the oldest temples is vested in the spirituality of their sites, in the form but

Figure 4.6 Rila Monastery, Bulgaria. The spiritual and symbolic associations of this site date back to the tenth century and the medieval monastery played an important part in Slav orthodoxy. Its inscription as a World Heritage Site in 1983 acknowledges its reconstruction, 1834–62, in the Bulgarian Renaissance style.

not the materials of their architecture, and where the buildings are periodically taken down and rebuilt. This is common practice for all types of structure in those many regions of the world where the principal building materials are intrinsically perishable.

Japan, which did not become a signatory to the *Convention* until 1992, symbolically hosted the Nara Conference on Authenticity in relation to the World Heritage Convention in 1994.

The resultant *Nara Document on Authenticity* challenged received thinking and paved the way generally for the World Heritage Committee to adopt a more inclusive approach to authenticity, one that respects cultural and heritage diversity around the world; recognises that dissimilar societies attach different sets of values to the original and subsequent characteristics of their cultural heritage; and enables the definition of authenticity to be assessed within each cultural context, not judged against others to which it may have no allegiance or connection (Figure 4.7).

The *Nara Document* proposes that assessments in any given instance should encompass matters relating to form and design; materials and substance; use and function; traditions and technologies; location and setting; and spirit and feeling. This represents a step-change from the uncomplicated

Figure 4.7 Oslo, Norway. The survival and conservation of perishable structures – wherever they occur in the world – requires approaches towards restoration that flout the strict concept of *authenticity*. Wooden churches are typical in many of the forested regions of Europe, from the Carpathian mountains in the east to Scandinavia in the north. The Gol stave church in the Norwegian Folk Museum, Oslo, dates from around 1200 and has twice been taken down and re-located. In the Bryggen quarter of Bergen, Norway (inscribed as a World Heritage Site in 1979) and Old Rauma, Finland (inscribed in 1991), the structures are almost entirely built of wood. Both historic areas have repeatedly been ravaged by fires; in neither case has periodic reconstruction been considered a determinant of loss of authenticity.

European-oriented definition that was quoted on page 8, namely, 'materially *original* or *genuine* as it was constructed and as it has aged and weathered in time'.

The *Nara Document* was conceived essentially as a route to embracing non-European cultural traditions into the World Heritage fold. It has however been explicitly embraced within Europe itself, and in a manner that questions the uncompromising heartland of established philosophy and practice in architectural conservation.

Carcassonne in France is primarily thought of – and marketed to tourists – as a medieval fortified city, complete with its castle, Gothic cathedral, and compact urban pattern. ICOMOS did not support the inclusion of Carcassonne on the World Heritage List when it was first nominated in the early-1980s. As with Rila Monastery, it did not – in their view at the time – satisfy the criterion of authenticity.

In 1849, Eugène Emmanuel Viollet-le-Duc, one of the rogues of the Anglo-French anti-restoration movement that led in England to the founding in 1877 of the Society for the Protection of Ancient Buildings, initiated a lengthy, thorough-going restoration programme that was not completed until 1910 (following Viollet-le-Duc's death in 1879, his pupil Paul Boeswillwald took charge of the works). These nineteenth-century interventions, which were strongly criticised during Viollet-le-Duc's lifetime for their lack of authenticity, had a determining effect on the Carcassonne we see today.

The nomination file was resubmitted in the mid-1990s and Carcassonne was inscribed as a World Heritage Site in 1997. This was not in spite of, but because of, the 'exceptional importance' of Viollet-le-Duc's restoration work 'which had a profound influence on subsequent developments in conservation principles and practice'. The new ICOMOS advisory report, written in the light of the 1994 *Nara Document*, describes the restoration as constituting 'a real element in the history of the town'. The report acknowledges that the stylistic restoration of Carcassonne challenges the philosophy and principles of the 1964 *Venice Charter*, but describes it as Viollet-le-Duc's 'master work'. The report recognises that our cultural heritage today owes much to the architect-conservators of the nineteenth century and quotes from the *Nara Document* as follows:

> It is . . . not possible to base judgements of values and authenticity within fixed criteria. On the contrary, the respect due to all cultures requires that heritage properties must [be] considered and judged within the cultural contexts to which they belong.

This debate advances the view that authenticity is not a restrictive concept either in time or space, and that just as each generation precedent to our own has contributed to the historical layers of the buildings and cities that they have inherited, so this and subsequent generations have an equally valid contribution to make – with the proviso that it is a positive and lasting one (Figure 4.8).

Figure 4.8 Durham, England. Durham Cathedral and Castle were inscribed as a World Heritage Site in 1986. Extensive reconstruction and restoration work carried out on the cathedral between 1780 and 1800 by the architect James Wyatt provoked an outcry at the time from the Society of Antiquaries. It was not considered a critical factor in the ICOMOS advisory report.

This debate also connects with the view that culture itself is not a restrictive concept but 'a process and a negotiation of connections'.

Integrity and historic cities

Related to authenticity is a second key concept against which World Heritage Sites are assessed, namely *integrity*. Integrity is defined in the UNESCO *Operational Guidelines* as 'a measure of ... wholeness and intactness', and in relation to historic cities there is an anticipation that the 'relationships and dynamic functions [that are] essential to their distinctive character should ... be maintained'.

Intangible cultural heritage

As well as *authenticity*, another shift in direction is the increasing importance that is attached to intangible cultural heritage. This is taken to encompass, amongst others, oral traditions and expressions, including language; performing arts; social practices, rituals and festive events; knowledge and practices

Figure 4.9 Etara, Bulgaria: continuity of expression in local musical traditions.

concerning nature and the universe; culinary arts; and traditional craftsmanship. Also, the consideration of these not simply as manifestations from the past that can be recorded, documented and pigeon-holed as *heritage*, but to the objective of securing their viability and creative continuity as an essential component of cultural diversity in today's and tomorrow's world (Figure 4.9).

Promotion of awareness and actions to protect the diversity of intangible cultural heritage form part of UNESCO's broadening representation of the linkages between culture and nature, cultural diversity and biodiversity, and their relationship to sustainability and sustainable development. Thus, in 2003, the *World Heritage Convention* was complemented by the *Convention for the Safeguarding of the Intangible Cultural Heritage*.

Again, as with the UN and UNESCO initiatives that could, mistakenly, be classified into *sustainability* and *conservation*, the compartments within UNESCO are not rigid, and the World Heritage Committee has been increasingly concerned to insinuate an anthropological approach into the World Heritage List, thereby underscoring further the relationship between heritage and society.

The Blaenavon Industrial Landscape was included on the World Heritage List in 2000 primarily in recognition of its place in the history of coal mining and iron and steel production in the first Industrial Revolution. Its inscription also recognises the distinctive working class culture that evolved throughout the nineteenth and early-twentieth centuries in South Wales, and which included education, non-conformist chapels, choirs, trade unions, and rugby.

Blaenavon is situated in an economically depressed part of the United Kingdom in a landscape that has been torn apart by its former industries.

Recognition, continuity, and extension of its community traditions are playing an important part in the revival of local pride in the area and in its socio-economic regeneration.

The World Heritage brand

Just as the *Montreal Protocol* has been a highly successful international environmental agreement emanating from the United Nations, so has the *World Heritage Convention* been a major instrument of international cooperation emanating from UNESCO.

The focus on World Heritage Sites – or, to be more precise, on getting on to the World Heritage List – is not, however, without its detractors. Is it, for example, an objective list of 'wonders of the world' or a competitive 'roll of honour'? Should Italy have the most sites because it has the most of a range of widely recognised cultural heritage from the Roman period onwards, or China because it has an equally rich but less well-known cultural heritage and is the most populous?

The emphasis of the World Heritage Committee in recent years has been on facilitating entry on to the list by first, establishing a global strategy that guides the categories that are most likely to be supported; and second, refining, clarifying and publishing detailed guidance on the procedures to be followed and the precise documentation to be provided for a successful nomination. Thereafter, especially in the fundamental issue of management, there is a guidance shortfall, and the gap between well-managed and unmanaged sites is substantial, even within individual state parties.

The focus on World Heritage Sites also risks the view that it is only sites that are, or have the prospect of becoming, recognised by the World Heritage brand that are important and whose protection and conservation should be supported.

In the United Kingdom, parliamentary and local authority groups representing World Heritage Sites lobby for priority to be given to them within protective legislation and funding regimes, thereby sidelining the overarching commitment made under Article 5 of the *Convention* and the related *Recommendation*, overlooking the significant promotional advantages which the brand brings, and ignoring the disparities within the list itself.

The World Heritage List includes, for example, a dozen historic cities in Italy (Figure 4.10) but only two in the United Kingdom: the city of Bath and the Old and New Towns of Edinburgh. Six tightly defined, linked areas of Liverpool are also included. Cities such as Cambridge, Chester, Norwich, Oxford and York might be thought of as potential candidates, but they do not fit into the post-1994 global strategy (Figure 4.11). Similar imbalances between the nations of Europe occur in every category of cultural heritage site: sites of antiquity; castles; cathedrals; country palaces and parks; bridges, canals and railways; industrial sites; and twentieth-century modernism.

Do such imbalances on an international list justify enhanced legal protection at a national level? And where is the World Heritage brand in relation

Figure 4.10 Venice, one of a dozen Italian cities that feature in the World Heritage List.

Figure 4.11 Cambridge, England, a United Kingdom city that might be thought of as a likely candidate for World Heritage Site status but which falls outside the post-1994 global strategy.

to the broader international agenda of sustainability if listed sites are to be subject to continuous prioritisation of external funding? Should not such sites aim to be models of sound management and of self-supporting conservation (a laudable objective in the 1968 Esher report for York), models that inform and guide the broader spectrum of a nation's cultural heritage – especially if credence is attached to the conclusion of the *Burra Charter*, namely, that 'the best conservation often involves the least work and can be inexpensive' (see page 14)?

As a conference leader asked recently: 'Do we need World Heritage Sites? . . . Should we not be focusing on the issues?'

Urban conservation: international cooperation

The *World Heritage Convention* is envisioned as a catalyst for international cooperation aimed at safeguarding and securing the long-term future of world cultural and natural heritage. The World Heritage List and the World Heritage List in Danger are the focus for this.

Vilnius, Lithuania

In 1990, Vilnius was propelled from the status of a regional centre at the western extremity of an empire that extended half way round the globe into that of the capital city of a sovereign state, and from a command to a free-market economy. As with the neighbouring Baltic States of Estonia and Latvia, Lithuania lacked recent history of an autonomous legal system or public administration. Incoming investors were attracted by the prospects of economic boom, and property speculation was rife.

Lithuania signed the *World Heritage Convention* in 1992 and Vilnius was included on the World Heritage List in 1994. This served to focus attention on the need to establish a framework within which the outstanding universal value of the site could be protected and conserved.

A revitalisation strategy was prepared in 1995–96, with support from the Lithuanian government, the municipality of Vilnius, UNESCO, and the World Bank, together with a team of experts from Scandinavian countries and from Edinburgh. This led in 1998 to the establishment of the Vilnius Old Town Renewal Agency (OTRA), an organisation whose vision and purpose are to establish an integrated approach to conservation-sensitive urban renewal in the city, coordinating public and private interests and initiatives (Figure 4.12).

OTRA is one of the few site-specific bodies that is dedicated to proactive urban conservation in a major city in Central and Eastern Europe. In the concept as well as in its operation, OTRA is regarded by ICCROM and others as a management model of importance not only for Vilnius but for other historic cities across the region. Achieving not-for-profit status in 2001, and supported by both national and municipal governments, it has established a multi-aspect programme of action.

Its wide-ranging objectives and actions include developing a database; coordinating infrastructure improvements and other revitalisation works funded by the state and the municipality; promoting and managing public

Figure 4.12 Vilnius, Lithuania, was inscribed as a World Heritage Site in 1994. The Old Town Renewal Agency was established in 1998 to promote and manage an integrated approach to conservation-sensitive urban renewal in the city and has been greatly assisted by support from international partners.

sector financial support to property owners in the Old Town; simplifying the approval procedures for conservation works and building permits; preparing investment studies and identifying regeneration opportunities; negotiating and mediating between different agencies and interests; opening an information centre; arranging exhibitions and seminars; raising awareness and involving the local community; contributing to the promotion of Vilnius as a visitor destination; and developing international cooperation (Figures 4.13 and 4.14).

OTRA has developed an impressive range of free publications, including *Conservation Guidelines*, all aimed at establishing common ownership amongst the local community, business sector, developers and other interests in the implementation of a conservation-sensitive vision for the city. Cultural events organised by others are contributing to raising the profile of the city's tangible and intangible heritage.

Operating within a challenging socio-economic environment, OTRA has sought to address a complex multi-faceted challenge in a coordinated way. It has been greatly assisted by programmes of training and capacity building that have been supported by a range of international partners.

Figures 4.13 and 4.14 Vilnius, Lithuania. The promotion of joint-proprietor projects forms a major part of the OTRA regeneration programme. A number of homeowners associations have been formed and are being grant-aided in the renovation of their tenements and associated courtyards, of which Vilnius boasts a wide variety of types and layouts.

A key problem that was identified on a management performance assignment early in 1999 was the absence of effective protective legislation. Indeed, an important historic building in the heart of the city centre had recently been demolished for redevelopment contrary to an approved scheme and the authorities were powerless to act. Once identified, and with full support from international partners, the problem was urgently addressed and new legislation put into place.

Dubrovnik, Croatia

Dubrovnik's recovery in recent decades from a succession of natural and manmade disasters is a singular tribute to the spirit of the local population, the solidarity of the international community, and the effectiveness of the *World Heritage Convention*.

Situated on the Dalmatian coast on the eastern seaboard of the Adriatic, Dubrovnik lies in a region of regular seismic activity. Historically, by far the most catastrophic earthquake occurred in 1667, destroying about half of the buildings within the city walls. It was the event that heralded the long, slow decline of this former city-state to a nadir in the 1930s and 1940s.

Figure 4.15 Dubrovnik, Croatia. Only about a quarter of the reconstruction work that was planned following the 1979 earthquake was completed prior to the 1991–92 siege. The 1979 earthquake caused serious damage to over a hundred monuments in the city and its immediate region and destroyed some areas of housing close to the sea walls (shown here). The reconstruction of these areas forms part of the ongoing programmes of the Institute for the Restoration of Dubrovnik.

The walled city of Dubrovnik was inscribed as a World Heritage Site in 1979. In the same year the city suffered another serious earthquake, one that severely damaged a number of key buildings, including the Rector's and Bishop's Palaces (see caption to Figure 1.8 on page 11), and destroyed some densely built areas of housing close to the sea walls (Figure 4.15). This immediately led to the founding of the Institute for the Restoration of Dubrovnik, which established an overall strategy and annual programmes of restoration. From the outset, these were supported by expertise and contributions from the international community.

During the seven-month siege of 1991–92, over 800 buildings within the old city received direct hits and several dozen were largely destroyed.

UNESCO placed Dubrovnik on the World Heritage List in Danger in 1991. Following the cessation of hostilities and in collaboration with the national and local institutions, UNESCO drew up an action plan, spearheaded an international appeal, and contributed technical as well as financial assistance.

Such was the success of the post-war restoration programme that Dubrovnik was removed from the World Heritage List in Danger in 1998. Furthermore, buildings damaged or destroyed in previous earthquakes are included in the ongoing programmes (Figures 4.16 and 4.17).

Dubrovnik and its surrounding archipelago suffered a severe recession in their tourist industry as a consequence of the 1991–92 siege and continuing conflict in neighbouring countries throughout the 1990s. As a result, the municipality has worked hard to achieve a more balanced year-round

Figure 4.16 The most obvious damage caused by the 1991–92 siege was the destruction of roof tiles throughout the old city. Initially, supplies from a variety of sources were used to patch and repair buildings. Concerned to achieve uniformity, the Institute for the Restoration of Dubrovnik negotiated a special design and supply with a local company, which it secures by guaranteeing a minimum annual order.

Figure 4.17 Funded in major part by the state, and in smaller shares by both the county and the municipality of Dubrovnik, the Institute for the Restoration of Dubrovnik has been responsible for major investment in the conservation of public spaces, religious and secular buildings, the city walls and fortifications, and the old harbour (shown here).

economy for the city as a whole by avoiding an excessive concentration of activities in the historic core and encouraging the development of non-tourist related functions, such as higher education. Dubrovnik University was formally established in 2003, and courses include media and communication, sea flora and fauna, and historic building conservation.

There has also been concern to maintain the viability of the walled city as a vibrant, mixed, local community that serves all ages (Figures 4.18 and 4.19). A new sports hall was completed in 2004, the primary, secondary and music schools within the city walls are the subject of substantial investment, and the municipality spends a substantial part of its annual budget on cultural activities for its citizens and for visitors, including the city's galleries, museums, theatres and orchestra. Additionally, the municipality has instigated a programme of 'Neighbour-to-Neighbour' campaigns that inform and involve citizens in caring for the city, its environment and its various communities.

International support for conservation work in the city has also benefited its surviving treasure chest of movable works of art. Many of the works, including paintings, sculptures, and gold plate, that were commissioned and collected under the Dubrovnik Republic were lost in the Great Earthquake of 1667 and the subsequent fire. Some of the most interesting and important works of the Dubrovnik School of Renaissance painters however survive, especially those of Lovro Dobričević (c.1420–78) and Nikola Božidarevič (1476–1517).

Figures 4.18 and 4.19 Dubrovnik, Croatia. Of the city-wide population of 45,000, 3,500 live in the walled city. Around sixty-five per cent of the available floorspace in the historic core is residential, and housing has been the immediate priority in successive waves of reconstruction and restoration. The community is mixed, including families with young children.

In the early-1990s, at a time when most attention was focused on the plight of the city's architecture, a conservation laboratory was inspired and equipped through the ARCH (Art Restoration for Cultural Heritage) Foundation, whose founder chairman, the Archduchess Francesca von Habsburg, also introduced restoration experts from abroad, especially from Italy, to provide training to local restorers. This work has highlighted the importance of the city's wider artistic inheritance, and is now incorporated into the state-funded Dubrovnik Conservation Department of the Croatian Conservation Institute.

In recent decades Dubrovnik has witnessed a succession of disasters and triumphs in the protection and conservation of its cultural heritage. The city's diverse tangible heritage – its architecture and its many artistic treasures – have all been supported through the international community, and the city's intangible cultural heritage features prominently in the programmes of successive mayors. This cultural heritage, and the relationships to the natural heritage of the surrounding inland and coastal region, are seen as central to community life in the city today (Figure 4.20).

Figure 4.20 Dubrovnik, Croatia. Market days in the Gundulićeva Poljana serve the local population and provide an outlet for producers from the surrounding islands and countryside.

Chapter 4: digest

Within the embracing UNESCO definition of heritage as 'our legacy from the past, what we live with today, and what we pass on to future generations', the 1972 *World Heritage Convention* and its accompanying *Recommendation . . . at National Level* are seminal documents that highlight the interdependence of cultural and natural heritage and the mutuality of their protection. Under the *Recommendation* we are urged to regard this heritage as a homogeneous whole, and under the *Convention* the category of *cultural landscape* recognises the 'combined works of nature and man'.

The *Recommendation* expands upon the importance of providing the cultural and natural heritage with an active function in today's communities, and integrating it into socio-economic and cultural life to the point that it is a determining factor in regional and national planning policy.

The 2003 *Convention for the Safeguarding of the Intangible Cultural Heritage* underscores the increasing emphasis on an anthropological approach to cultural heritage and the importance that is attached to creative continuity as an essential component of cultural diversity in the world.

The relationship between place and people is expressed through the definition of *integrity* in relation to historic cities, which seeks to safeguard functional as well as artistic and material continuity. The accepted definition of *authenticity* as 'materially *original* . . . ' is challenged by the *Dresden* and *Nara Declarations* to encompass both reconstructions and stylistic restorations. Between them, these declarations recognise spiritual as well as material values, continuity of historical layers, and extend the previously restrictive concept of *authenticity* both in time and space.

UNESCO provides an important platform for international support and cooperation to the benefit of sites that are inscribed on the World Heritage List, but the focus on entry on to that list has over-shadowed the commitment made by state parties to the protection and conservation of national heritage, and the emphasis in the United Kingdom on interpretation reinforces a limited definition of *heritage* that separates it from everyday life.

Vilnius and Dubrovnik offer good examples of coordinating efforts between the public and private sectors and between international organisations and individuals to protect, conserve and secure a pivotal functioning role for cultural heritage in the life of communities, even in the most challenging of situations.

Chapter 5
Conservation: United Kingdom Position and Directions

Established protective system

English Heritage, like its counterparts Historic Scotland and Cadw (Wales's historic environment division), is a discrete governmental organisation that acts as statutory advisor to other sectors of government on conservation issues and policy. It is responsible for promoting the conservation and appreciation of the built heritage, including through the publication of guidance and as a provider of education; guardianship and management of public access to state-owned monuments and sites; the identification of historic structures and sites for statutory protection and overseeing the various mechanisms of control; and managing programmes of grant aid.

From the first Ancient Monuments Act of 1882, through the 1944 Town and Country Planning Act, to the 1967 Civic Amenities Act and their various and numerous revisions and regional adjustments, ancient monuments (essentially non-inhabited structures and remains including ruins and archaeological sites), historic buildings (essentially habitable structures) and conservation areas have been dealt with under separate legislation and from different perspectives.

Inevitably, this poses anomalies. The distinction, for example, between an ancient monument that was once inhabited and has become a ruin – such as a castle – and a historic building that is in principle habitable but which, whether through planning blight or neglect, is also manifestly a ruin, is a judgement based on history, condition and expectation. It is not so much a question of whether the structure has a roof on it – though that clearly is a factor that relates to its condition – more a question of how long it has been a ruin and, critically, whether the archaeological–historical view is that it should be preserved as such.

The anomaly has been resolved in some instances by classifying a ruin both as an ancient monument and as a historic building, an expediency that – on the face of it – keeps the options open, but which in practice means that a private owner has two parallel sets of bureaucracy to deal with even over matters of maintenance (Figures 5.1 and 5.2). This distinction between

Fatlips Castle: *exterior*.   Fatlips Castle: *interior*.

Figures 5.1 and 5.2 Fatlips Castle, Scotland. This ruinous sixteenth-century tower house, situated near Minto in the Scottish Borders, is both a scheduled ancient monument and a listed building. Action to secure its future has been inhibited by the fact that it falls between two sets of protective legislation, each of which has a different objective.

ancient monuments and historic buildings owes much to the eighteenth-century Picturesque Movement and to the influences that inspired the *SPAB Manifesto* of 1877, which predated the first Ancient Monuments Act by a mere five years. The distinction in protective status between non-inhabited and habitable structures is not one that I have encountered outside the United Kingdom.

The United Kingdom's protective system is, statistically by population, one of the most inclusive in the world. At July 2006 it embraced almost 36,000 ancient monuments, over half-a-million historic buildings, and around 10,500 conservation areas. This constitutes a significant component of the nation's built inheritance, from the vernacular to the monumental.

Estimates published by English Heritage in 2000 suggested that around five per cent of the building stock in England at the time was listed and around ten per cent fell within conservation areas; additionally, that nearly half of the housing in England was over fifty years old and a quarter dated from before the First World War. This, together with its equivalent in non-residential use, constitutes an inheritance that is as permanent as the materials of which it is constructed and as society chooses to maintain and make use of it.

The involvement of civil society, technical and policy guidance

Apart from the legislative position, the United Kingdom has long been noted for the wide-ranging involvement of civil society in the protection and conservation of historic buildings and sites, especially through amenity societies and charitable trusts.

The Society for the Protection of Ancient Buildings, founded in 1877, was one of the earliest of many hundreds of national and local societies that represent different historical and architectural periods, buildings types, geographical areas and special interests. Many of these are grouped under the Civic Trust (England and Wales; founded in 1957) and the Scottish Civic Trust (founded in 1967).

The National Trust, founded in 1894 and with a membership today of 3.4 million, has long been the largest landowner in England, Wales and Northern Ireland, and manages directly or indirectly several hundred stately homes, parks, landscapes and other properties, almost all of which are open to public access (Figures 5.3 and 5.4).

Its counterpart, the National Trust for Scotland, founded in 1931 and with a membership of around 300,000, has also made a significant contribution to the survival and conservation of vernacular architecture through its Little Houses Improvement Scheme – which was one of the United Kingdom's four showcase projects in European Architectural Heritage Year 1975. Initially focused on fishing villages in the county of Fife, the scheme has been extended to other parts of Scotland (Figures 5.5 and 5.6).

Additionally, across the United Kingdom there are over 200 building preservation trusts that acquire, restore and find new uses for historic buildings, many of them on a revolving fund basis.

One might assume, with the inclusive nature of the legal protection and the degree of involvement by civil society, that the United Kingdom is imbued with a conservation ethic and that its technical, professional and management skills are amongst the best in the world.

Certainly, on the technical side, the profusion of textbooks – from the 1920s onwards – and the abundance of other publications and leaflets on every aspect of building and historic area conservation, including policy guidance from government, is impressive.

Equally, the degree of public interest is wide-ranging, and the popularity of television programmes such as *Restoration* bears witness to the enthusiasm for the rescue and restoration of individual historic buildings (Figures 5.7 and 5.8).

Shugborough Hall: *entrance facade.*

Figures 5.3 and 5.4 Shugborough Hall, England, the ancestral home of the Earls of Lichfield, is owned by the National Trust and managed by Staffordshire County Council as a working historic estate, complete with costumed guides and the sights, smells and tastes of the 1800s. An important visitor attraction, it symbolises how many people identify with *heritage*.

Shugborough Hall: *ornamental garden.*

Figure 5.5 Pittenweem, Scotland. The National Trust for Scotland's pioneering programmes aimed at saving many of Scotland's smaller historic houses began in the early-1930s in the burgh of Culross, and led to the conversion into modern homes of structures that would otherwise have been demolished as slums. The Gyles, Pittenweem (centre of picture) was restored and converted into four dwellings. The project won a Civic Trust award in 1965 (Wheeler & Sproson, architects). Gyles House (right of picture), which had already been restored, was acquired by the Trust in 1972.

Figure 5.6 South-East Scotland. The author's house was acquired through the National Trust for Scotland's Little Houses Improvement Scheme and restored under a conservation agreement (restored 1980–83; Dennis Rodwell, architect).

Britannia Music Hall: *street elevation to Trongate.* Britannia Music Hall: *interior of auditorium.*

Figures 5.7 and 5.8 Glasgow, Scotland. The Britannia Music Hall, which dates from 1857, is reputedly the first music hall to have been built in Scotland and is one of the oldest in Britain. As a place of popular entertainment it operated until 1938, since which time is has survived as though in a time capsule. Entertainers such as Stan Laurel – of Laurel and Hardy fame – began their careers at the Britannia. The Britannia featured on the BBC2 programme *Restoration* in 2003. The concept for its restoration has been a project of the Glasgow Building Preservation Trust. (Feasibility study 1993; Dennis Rodwell, architect.)

Failures within the established protective system

There are, however, certain matters that may be considered to represent key failures within the established protective system.

Administered locally, the historic building and conservation area legislation often falls short of its objectives through a combination of lack of political will and direction by management. This affects decision making at a policy level and results at a practical level. It impacts on the public perception of what conservation professionals seek to achieve and how society reacts to them.

An important example of this concerns the United Kingdom's 10,500 conservation areas. On the premise of their architectural and historic interest, these are designated by local authorities under legislation that exhorts the preservation and enhancement of their character and appearance. Yet in

Architectural features and local distinctiveness

The village settlement, Darley Abbey, Derby, England

Architectural features give tangible expression to geocultural identity.

The factory village of Darley Abbey was inscribed as part of the Derwent Valley Mills World Heritage Site in 2001. Whereas the industrial roots of the village date back to the medieval monastic period, the principal pioneering developments in its cotton manufacturing industry and the associated settlement of workers' housing date from the 1780s and were completed by the 1880s (Figure 5.9).

Figure 5.9 Darley Street comprises a mixture of house shapes and sizes, certain of which pre-date the establishment of the cotton industry in the 1780s. For a number of years the workers' housing was under the threat of slum clearance. This was lifted in the early-1970s by the designation of the village as a conservation area and reinforced by its declaration as a general improvement area.

The former cotton mills and the village settlement are located on opposite sides of the river Derwent. Analysis of the two sets of buildings discloses a progression of architectural styles and features that is unique to their purpose and to their date of construction. Each and every industrial and domestic building or group was built to a different architectural vocabulary. Within very short times spans, the differences are sometimes marginal; within longer spans, considerable (Figures 5.10–5.13).

Despite an overall development period of only a hundred years and the particularity that it was built as a discrete settlement under the ownership of a single family, Darley Abbey is a microcosm of the characteristics that contribute to the local distinctiveness of any urbanised area.

Left: The terrace in Brick Row was constructed 1797–1800 (see also Figure 2.5 at page 27), and part of the top floor was initially used to accommodate schoolrooms for the village children. Some of the original multi-paned timber windows with cast iron opening casements survive.

Figures 5.10 and 5.11 (above and right): Windows, and **Figures 5.12 and 5.13 (below and below right):** Doors.

Above: Nos 3 to 9 New Road were built in 1826. The first floor windows were of an unusual pattern that may have been unique in the village: the upper half is hinged from the top, the lower half is fixed. A few examples survive at the rear elevation.

Poplar Row was completed by 1823. This is a rare surviving original example of the pattern of planked doors that was typical of the earlier workers' housing throughout the village.

The purpose-built school in Brick Row was constructed in 1826. It exhibits mature neoclassical construction and detailing in the Georgian style, and its original features, which include stone, brick, joinery and ironwork, are very substantially complete.

Many of the original architectural features in the village settlement survive; others were modified in the nineteenth century; a significant number were lost during the period of grant-aided housing improvements in the 1970s (Figure 5.14); and the reinstatement of some has since been supported by conservation area and historic buildings' grant aid.

Figure 5.14 Darley Abbey was inscribed as part of the Derwent Valley Mills World Heritage Site in 2001. Completed by 1792, the Four Houses are amongst the most historically important of the workers' housing in the industrial village. It is the earliest known example of a 'cluster house': a detached group of four houses constructed in a single block. Originally finished in brickwork, and fitted with doors and windows that have since been changed, the building was the subject of grant-aided housing improvements in the 1970s. Conservation area controls at the time were insufficient to prevent significant alterations to the block, such that it was not considered to merit statutory protection under an English Heritage listing review in 2002.

The features that survive testify to the criteria of authenticity and integrity that are anticipated within the relevant parameters of the *World Heritage Convention* and the *Operational Guidelines*.

There is substantive evidence that the survival of historical features supports the market value of properties far better than their loss.

only one out of ten of these conservation areas have those same local authorities put in place the set of regulations (directions) that enable control of the architectural features and finishes of the generality of the buildings, aspects that government guidance, practising conservation professionals, and many others insist define those areas' character and appearance.

It is a debate that relates to all of the numerous aspects of local distinctiveness that are expressed through the buildings of any one historic area in comparison to another. It concerns changes to wall and roof finishes, chimneys, porches, garden walls and railings, paint colours, shop fronts, signage and advertising, all of which traditionally were unique to each and every closely defined geographical area, neighbourhood, or even street, and which expressed in architectural and townscape terms the identity of that place and its community. A particular focus of change has been on traditional external joinery – especially windows and doors – where society generally has been persuaded by marketing and fashion that comparatively short-life plastic replacements represent an improvement and are more energy efficient.

The issue has been a bone of contention ever since the passing of the Civic Amenities Act in 1967 and remains unresolved. The reform of the delegated system of regulation and its replacement by nationwide regulations withdrawing 'permitted development rights' in conservation areas was forcibly represented in 1992 by the English Historic Towns Forum in *Townscape in Trouble: Conservation Areas – The Case for Change*. In 2000 – and 'as an immediate priority' – a similar call was made in *Power of Place* (see below). As *Townscape in Trouble* states: 'There are ... more planning controls over the appearance of an inner city tower block than a house in a conservation area'.

The absence of consistent policy in conservation areas – which has certain parallels with practice relating to the more than ninety per cent of listed buildings that do not fall into the top categories – has had several knock-on effects.

First, it has marginalised those – especially conservation officers working in local government, who may be working in isolation and are almost invariably operating below management level – who seek to abide by what central government and their peers insist is good conservation area practice. An English Heritage publication of 2000 summarises the position more forcibly: 'The inadequate supply of suitably qualified and experienced staff, particularly in local government, is ... a matter for concern, as is the positioning of such specialists within organisations at levels where they can exert too little influence on decision-making'.

Second, in the forty years since the passing of the Civic Amenities Act, the scale of the changes that have taken place across the majority of the 10,500 conservation areas is such that a large number of them have lost the architectural and historic interest for which they were designated and have little left to preserve and enhance. Their continued designation as conservation areas is largely meaningless. There may be other sound reasons for protecting them – such as the ambiguous presumption against demolition – but the essential premiss under existing legislation is open to challenge.

Figure 5.15 Edinburgh, Scotland. The supply and demand equation can tip the balance in favour of traditional materials and skills. In Edinburgh there have been periods when the demand for conservation works, matched by the availability of materials and craft skills, has led to the use of natural stone and traditional joinery being on a par with or less expensive than artificial alternatives such as reconstituted stone or plastic. This results from consistency in the application of conservation policies towards historic buildings and areas. (Signal Tower, Leith, Edinburgh. Constructed originally as a windmill in 1685–86; Robert Mylne, master mason. Converted to a watchtower at the time of the Napoleonic Wars; converted again and extended c.1810 to form a tenement of eight flats. Tercentenary restoration, 1985–86; Dennis Rodwell, architect.)

Third, the lack of coordinated policy has severely limited the demand for and hence the availability of traditional craft skills, and where those skills can be found they have become specialised and relatively expensive. It is the basic law of supply and demand in a free-market economy.

The sanitisation of historic city centres – such as promoted in the Esher report, adopted in York and repeated in other cities – has additionally had the effect of relocating traditional craft industries away from their market places. For craft skills to survive and to have a sufficient workload to be efficient and competitive they must have work and be located where that work is (Figure 5.15). The emerging economies of Central and Eastern Europe understand this and are acting accordingly (Figure 5.16).

In conservation areas especially, and from a policy through to a practical level, the picture is one of lack of direction, confusion and frustration. When, for example, architects and building contractors are undertaking work within the remits of a number of different local authorities – which is the norm – they will be dealing with a different conservation culture in each. The resultant, inevitable aggravation impacts on society's perceptions. If one person living in a conservation area within one local authority's jurisdiction is allowed to do one thing, and her or his friend or relative in another is

Figure 5.16 Kutna Horá, Czech Republic. The urban renewal programmes in historic cities of the Czech Republic have included support for local craft businesses. A stone carver occupies premises at the end of this street, which is situated a short distance from the market square.

required to do something else at increased cost, then the second will not be a natural supporter of conservation. The media likes aggrieved people – especially in such instances, local newspapers and radio – and politicians depend on their support at elections.

Another key issue, one that has been contested since the early-1970s, concerns the discrimination between repair work on the one hand and new construction work on the other in the liability for value added tax (VAT). If VAT is applied (currently at 17.5 per cent) on repair work and at zero on new work, it is a burden on owners who seek to maintain their properties and a signal from government of encouragement to alterations or redevelopment of the kind that opposes the conservation approach that the protective legislation and guidance expects. In addition, the United Kingdom is the only nation in Europe not to allow taxation relief for the maintenance of historic property.

To be concise if harsh, there are aspects of the established protective system coupled with fiscal discouragement that render it barely fit for purpose. There is a minefield of contradictions in the system's operation, and the cumulative degradation in recent decades in core values – especially local distinctiveness – across the majority of the United Kingdom's supposedly protected historic areas is very substantial.

Such – from an architectural conservation perspective – is part of the starting point for any review of the protective system.

Review of policies relating to the historic environment in England

Initiation

Early in the year 2000 English Heritage issued a consultative document inviting participation in a comprehensive review of policies for the historic environment, thereby initiating a process that is not expected to reach a clear conclusion in implementation for several years.

The use of the embracing term *historic environment* – defined broadly as 'all the physical evidence for past human activity, and its associations, that people can see, understand and feel in the present world' – was significant. It placed the review into a context in which the hitherto disparate elements (which, apart from the three components of the built heritage, also include parks and gardens, and battlefields) were being evoked as components of a broader whole. Indeed, a whole that was potentially far greater than the sum of its parts: selective, but without chronological, thematic, geographic, cultural or ethnic limits, and regardless of scale – from the locally distinctive through the internationally significant to the entirety of the historic landscape.

This offered the opportunity to broaden the perception of the built heritage beyond the limited notion of architectural and historic interest, to strengthen the value of conservation to incorporate resource and societal values, subsume social inclusion, cultural diversity and community involvement, and to integrate policies for the built heritage with wider environmental concerns, including policies relating to the natural heritage. This would then ally the protection of the historic environment with the maturing international agenda of sustainability, and situate its conservation as a coordinated mainstream activity rather than a specialism, within central and local government, the professions and the construction industry, and in the public mind.

Of course, in order to achieve common ownership of a conservation-oriented ethic, there would need to be a two-way exercise in which the archaeologist–historian bias became more inclusive in its definitions and embraced wider socio-economic objectives. Reciprocally, those many who perceived the historic environment as a financial burden, an impediment to change, and of value primarily as a museum archive and raw material for the tourist industry, would need to acknowledge and affirm the historic environment as something of direct relevance and benefit to their daily lives, not least, socially and economically.

On a piecemeal basis, substantial strides had already been made along this path, but the opportunity for a major, mutual change in approach was especially ripe in relation to historic cities, which have the highest concentrations of historic buildings and conservation areas and which, in a holistic sense, were excluded from the established protective system. The challenge would be to implement in the most populous part of the United Kingdom precisely the kind of integrated approach to urban conservation that was

Figure 5.17 Kutna Horá, Czech Republic. In historic cities in the Czech Republic, including Český Krumlov, Telč, and Kutna Horá, integrated conservation on the model of the 1975 *European Charter* is the norm. In Kutna Horá, for example, underground service infrastructure renewal was coordinated with a traffic management scheme and the renewal of hard and soft landscaping – including in the market square, shown here. The viability of the local community was secured through advice and support to residents in the improvement and restoration of their houses coupled with the long-term allocation of commercial premises to small businesses that both employ and serve the local population. The tourism marketing strategy is coordinated with the many cultural sites and recreational opportunities in the neighbourhood of the city and is directed towards attracting visitors who will stay for several days and make beneficial use of local services.

espoused in the 1975 *European Charter* (see pages 12 to 13), in which the historic environment was insinuated on equal terms into the general planning framework, including policies on transport, land use, development priorities, and housing, and in which it would serve as the stimulus for balanced, sustainable regeneration (Figure 5.17).

Of direct relevance in the year 2000 was the obvious potential linkage with another widely trumpeted government initiative of the time, the Urban Task Force, whose report had recently been published (see Chapter 6).

Representation on this and related themes – including the failings in the established protective system noted above – was made at the time.

Power of Place

The first key step in this policy review was the publication in late-2000 of *Power of Place: The Future of the Historic Environment*.

This report was significant in drawing attention, through the interpretation of the results of a survey of public opinion, to widespread public support for maintaining protection in the historic environment: including for its value in defining the country's character of place and the identity of its people. A parallel survey, however, indicated that over three quarters of those polled did not identify the heritage as 'who we are/part of our identity'.

Power of Place recognised that the historic environment is seen by most people as a totality and that people value places as a whole, not just as a series of individual sites and buildings. It did not however explore 'power of place' from wider environmental and societal perspectives.

Within its terms of reference, *Power of Place* expressed the need to move on from a piecemeal approach to designations to a more holistic understanding and coordinated approach to conservation, one in which character appraisals (evolved from townscape appraisals) are promoted as the key tool for the management of change, and in a climate in which 'the historic environment is the context within which new development happens'.

The report nevertheless pre-supposed that the historic environment accorded with the established conservation approach based on selectivity and value judgements about cultural significance within academic criteria. And although the *environmental capital* (the embodied materials and energy) of existing buildings was recognised, the report did not explain how this fits in with an emphasis on selective value judgements and on new development. The report reads therefore as though it is only the environmental capital of structures that are assessed to have cultural significance that count.

Power of Place expressed the importance of the historic environment as an educational resource about the past as well as for its contribution to cultural life today, but did not explore what the public understands by conservation and the extent to which it is in tune with established practice; especially where it impinges on people's everyday lives within local communities and on the extent to which it is or is not succeeding in sustaining character and identity – at the local as well as at the national level.

The report underscored the importance of preventative maintenance, and although it made a case for training in craft skills it did not relate this to the existence of an equivalent level of demand. It also made an important call for the training of conservation officers to be broadened to incorporate knowledge of the property market and of the historic landscape. Additionally, the report called for a greater shared understanding of the historic environment and urban design issues between archaeologists, architects, conservation and planning officers, and surveyors.

Power of Place emphasised the importance of the historic environment to the country's tourist industry and the need for sustainable visitor management. The report also promoted the importance of 'creating the heritage of the future' by good design in new architecture – a call that may be interpreted as an encouragement to iconic modern buildings of the kind that do not necessary fit comfortably within the characterisation methodology.

Operating within the limits of a thesis in which *heritage* is synonymous with *history*, and in which character is a concept that focuses on an urban design approach to managing change rather than the relationship between the identity of place and community, *Power of Place* was a keynote document of its time and place. It made an important call for improved leadership in the sector at all levels, and for the responsibilities of 'green ministers' to encompass the historic environment.

Apart from strong advocacy of conservation-led regeneration, for which it cited examples targeted on individual properties and specific locations, *Power of Place* did not put forward suggestions for an integrated approach to urban conservation, an approach that would help obviate one of the problem areas that the report identified: vacant and underused historic buildings in cities (Figure 5.18). Nor, indeed, did *Power of Place* establish any substantive linkages with the Urban Task Force other than passing mention of it. Given that the preface confirms that this policy review is a once-in-a-generation opportunity, *Power of Place* appeared as a missed opportunity for the conservation interest to make a serious impact in historic cities.

Recognition of wider international agendas and initiatives was absent from the report and, as was pointed out by the Local Authorities World Heritage Forum in its formal response to *Power of Place*, although the United Kingdom signed the *World Heritage Convention* in 1984, it has yet to set out how it intends to discharge its obligations under it.

Figure 5.18 Derby, England. The disuse of upper floors is a major unresolved issue in historic cities across England.

Taking the policy review forward

Power of Place was followed in 2001 by the publication by government of *The Historic Environment: A Force for Our Future*.

A Force for Our Future was primarily significant for the commitment it made on behalf of government to reviewing the case for an integrated protective regime, a review that was formally announced in late-2002. The report also promised cross-departmental leadership at governmental level, and recommended the appointment of 'heritage champions' at local level.

A Force for Our Future reported that over half the annual turnover of the construction industry relates to repairs and maintenance. It also made an important connection at a theoretical level between the demand for traditional building crafts and the Urban Task Force, whose report *Towards an Urban Renaissance*, published in 1999, had identified a high level of medium-term stability in the urban fabric of cities. To be maintained economically and to a conservation standard would call for the availability of a significant number and range of traditional craft skills. In the absence of an effective regulatory framework, however, the theoretical connection lacks the foundation for translation into general practice.

Figure 5.19 Derby, England. The former main post office was the subject in the early-2000s of repeated listed building consent applications for the conversion of its lower floors to a bar and restaurant; as with other similar development proposals in the city, use of the upper floors was not incorporated into the schemes. The proposals were highly destructive to the internal structure and layout, but attempts to justify them relied on the sum of money that would be invested in the building. The amount of financial investment is no guarantee of necessary or wise expenditure either on historic buildings or in historic areas.

One of the central themes of *A Force for Our Future* is unlocking the full potential of England's historic environment to contribute to the quality of life. Emphasis is placed on the community benefits of the historic environment – local distinctiveness contributing to a sense of identity and belonging – and the importance of broadening intellectual and physical access to include all sections and age groups.

A Force for Our Future affirmed the economic benefits of the historic environment through tourism and as a catalyst for regeneration as well as its role as a stimulus for creative new architecture. Although the high-profile culturally led regeneration of the Albert Dock in Liverpool – with a cultural range that includes the Merseyside Maritime Museum, Tate Liverpool and the Beatles Story – is cited, the benefits of regeneration are principally equated in financial terms: to the amount of investment activity it generates, not on any inherent contribution to sustainability (Figure 5.19). Indeed, although both *sustainability* and *sustainable development* are referred to, the perception of how the historic environment is situated within the broader agenda of sustainability is ambiguous, with the important exception that it is recognised as a fragile, precious and non-renewable resource.

Conclusion of the policy review in England

The consultation paper *Protecting Our Historic Environment: Making the System Work Better*, published in 2003, laid the foundations for the government's proposals for improving the protective system, and these were set out a year later in the report *Review of Heritage Protection: The Way Forward*.

The Way Forward summarised the aims of the review as the delivery of 'a positive approach to managing the historic environment which would be ... central to social, environmental and economic agendas at a *local and community* as well as national level; and an historic environment legislative framework that provided for the management and enabling of change rather than its prevention'. Included among the objectives is to 'design an effective designation and control system ... [that] would protect and sustain the historic environment as a whole as well as its constituent parts and *would put the historic environment at the heart of the community*'.

Central to the outcomes of the review is the proposal to establish a new unified Register of Historic Sites and Buildings in England within an overarching definition of 'heritage assets'. This single register would be in two parts: one encompassing national designations; the other encompassing local designations, namely, conservation areas and locally listed items. All sites and structures covered by existing designations will automatically be transferred on to the new register.

It is envisaged that the new national register will be divided into three categories: above- and below-ground archaeological and monumental sites; historic buildings and structures in or assessed as suitable for use; and historic land and seascapes. It would also include World Heritage Sites, but

English Heritage as a regeneration agency

The shift in emphasis within English Heritage from the focus on conservation in the established sense – encapsulated generally as minimum intervention and defined in the *Burra Charter* as 'all the processes of looking after a place so as to retain its cultural significance' – is summed up in the new vision for conservation set out by English Heritage in 2000, where it is defined as 'the dynamic process of managing change'.

In *Conservation-led Regeneration*, published in 1998, English Heritage describes itself as a regeneration agency. In so doing it highlights its role in introducing Conservation Area Partnership schemes (CAPs) and Heritage Economic Regeneration Schemes (HERS) as well as, previously, Town Schemes; and alongside HERS its important advisory role through projects and area schemes supported by the Heritage Lottery Fund, including the Townscape Heritage Initiative.

These and related grant-aided schemes have and continue to make a major beneficial impact on the survival of individual buildings and the revival of targeted neighbourhoods, especially, more recently, in less-favoured areas (Figure 5.20).

Figure 5.20 Saltaire, Bradford, England. English Heritage and Bradford Metropolitan Council made a substantial contribution to the funding that spearheaded the regeneration of Salts Mill and the adjoining model industrial village in the 1990s. Saltaire, built 1851–76 by Sir Titus Salt beside the river Aire, was placed on the World Heritage List in 2001.

The cautionary note that needs to be sounded is simple. A significant proportion of the projects supported under these schemes result from some form of planning blight caused by a distortion of view and values attached to the inner parts of historic cities – whether through central area redevelopment, slum clearance, strategies of separated land use, or other. Additionally, a not insignificant component of the public funds that have been and continue to be invested into these schemes is undoing or redoing works that have either been supported previously through house improvement grants or result from the absence or lax enforcement of conservation area controls.

For the multiplicity of current regeneration-related schemes to be sustainable in the long term, they need to be supported by an integrated approach to historic cities that secures their socio-economic and environmental conditions.

> Let us be clear, the urban problems of our cities, towns and villages will not alone be solved by flinging money at them. (Sir Jocelyn Stevens, Chairman of English Heritage; quoted in *Conservation-led Regeneration*, 1998)

there are no current proposals to introduce additional controls in relation to them. The established underlying distinction between ancient monuments and historic buildings would be maintained, but there would be a new integrated consent regime administered at local authority level.

Amongst the most useful elements of the revised system is the proposal to prepare and publish 'statements of importance' for properties on the register – a proposal that has already been implemented by English Heritage for new designations under the established system. These set out what should be protected and why, and offer clarity to those responsible for caring for protected properties.

A complementary proposal to offer the facility of statutory management agreements would place complex sites outside the normal administrative consent process. This will be of particular benefit to the likes of a country estate, complete with its mansion house, outbuildings, ornamental and walled gardens, a landscaped park, preserved ruins and estate village, which fall within four separate designations under the established system. World Heritage Sites are already subject to management plans.

Whereas there is a recognition that the system of conservation areas is failing in numerous ways and in many locations, there is no commitment to introduce a blanket withdrawal of permitted development rights; rather, there is an expectation that local authorities will devote considerably more staff resources to the preparation of conservation area appraisals and management plans.

Questions

There are a number of issues that remain unclear from the policy review, of which three of the most obvious are:

First, how do they relate coherently to complex, larger-scale environments such as historic cities and to the many statements of recognition of the major contribution that the historic environment can make to urban regeneration and the creation of sustainable communities? None of the heritage assets selected by English Heritage as test-bed cases for the reformed system includes a complex, multi-ownership urban area where national designations such as historic buildings overlap local designations such as conservation areas.

Second, where and how are the additional professional skills to be secured and resourced at local authority level both to manage the new integrated consent regime and to deal effectively and proactively with conservation area issues?

Third, how is the recognised need for the increased training and availability of skills in traditional building crafts to be translated into a comparable demand for their services, and where are they to be located?

Urban conservation: a cathedral city

Chartres, France

Dame Jennifer Jenkins, formerly chairman of the National Trust, is quoted in *Townscape in Trouble* as follows: 'Britain's record in protecting historic areas is in my view less satisfactory than that of France, Italy or Holland.' Alongside, is a photograph of the Rue des Ecuyers in the lower town of Chartres (Figures 5.21 and 5.22).

Figure 5.21 Chartres, France. View across the river Eure to the lower town and the cathedral beyond. Chartres cathedral was inscribed on the World Heritage List in 1979.

Figure 5.22 Chartres, France. The Rue des Ecuyers after completion of the restoration works under the *secteur sauvegardé* action area (in 1975). The buildings date from the fifteenth century and include several early examples that are half-timbered and others of later date constructed of stone.

Declared a *secteur sauvegardé* under the 1962 *Loi Malraux*, the 64 hectares within the former medieval city walls comprise a number of distinctive quarters, including the cathedral and its precinct, and the upper and lower towns.

The strategy for the *secteur sauvegardé* and its relationship to the wider city has included the denial of through traffic; a combination of pedestrianisation and pedestrian priority; a set of three vehicle access loops into the *secteur* linked by the nineteenth-century boulevards, which have not been upgraded as a modern inner ring road; and the focusing of small-scale commercial uses – businesses, shops and leisure activities – in the upper town, balanced by larger-scale edge-of-city retail and associated parks that serve the city-region. The compact and densely built upper town also retains a significant residential component of houses and flats, and continuity of this has been secured.

The issue of city centre versus out-of-town shopping centres has been hotly debated in the United Kingdom. Current policy reinforces the focusing of retail uses and associated development pressures in city centres.

In Chartres, the strategic approach has benefited the historic centre in several ways. It has discouraged pressures for redevelopment; sustained a balanced city centre community; retained the traditional diversity of small-scale retail and other commercial outlets; and encouraged people to spend time in the city centre.

The lower town borders the river Eure. As a mixed residential and artisan quarter it was the focus of craft industries in the city. By the 1950s, the lower town had fallen into serious disrepair: its industries had all but disappeared, many of its buildings had been abandoned, and others had been undermined by the flow of the river and collapsed.

To kick-start the regeneration programme, a three hectare action area was designated, focused on the Rue des Ecuyers and its neighbouring streets. Working with the existing owners and supported by financial incentives, the surviving houses were restored – as a mixture of houses and flats.

Additionally, plans were prepared for the construction of new buildings on the vacant plots, all with the view to re-establishing a balanced community of all ages, including the provision of sheltered housing in newly built flats overlooking the river Eure. Small shops and restaurants have also been established in the area.

The principal scheme of restoration to the older houses in the action area was completed by 1974. Such was its impact in restoring confidence in the lower town that owners in neighbouring sections were – as hoped for and anticipated – inspired to upgrade and restore their properties without financial support.

The concept behind the new developments has been to integrate modern buildings into the historic street pattern and overall setting, and to apply plot ratios and adopt forms that contribute continuity to the mixed architectural vocabulary of the area (Figures 5.23 and 5.24).

The approach at Chartres is typical of that adopted in other historic cities across France: initially, spearheaded by the interventionist conservation planning and financial mechanisms invoked through the *Loi Malraux*; later, extended by example and market confidence through general planning policy.

The balance of commercial viability between town centre and out-of-town retail is the subject of ongoing debate in specific locations, but the overview is that the strategy is correct and that the long-term viability of historic cities as balanced, self-sustaining communities retaining a substantial residential component has been strengthened. As the caption to a cartoon published in 2000 noted: 'If historic buildings are really so valuable, how come no one is building any any more?'

Figure 5.23 Chartres, France. Old and new buildings either side of the river Eure.

Figure 5.24 Chartres, France. Modern housing on vacant plots in the lower town, designed to contribute continuity to the mixed architectural vocabulary of the area.

Chapter 5: digest

The United Kingdom protective system is, on the face of it, one of the most inclusive in the world and embraces a significant component of the nation's built inheritance at all scales from the vernacular to the monumental. It is a system that is supported by an abundance of technical advice and policy guidance.

Additionally, the United Kingdom is renowned for the wide-ranging involvement of civil society in protection and conservation activities, especially through amenity societies and charitable trusts.

The established protective system does not address historic cities holistically. Although it exhorts local authorities to designate conservation areas, there is an absence of nationwide controls that safeguard the characteristics which define – in a physical sense – their local distinctiveness, and that offer the potential to contribute to societal agendas of sense of place, community belonging, and social inclusion.

The lack of consistency at policy and practical levels has contributed substantially to the loss of authenticity and integrity in historic areas in Britain. It has also led to a reduction in demand for the traditional craft skills on which the sound repair and maintenance of historic buildings and cities depends. This in turn has led to an unfavourable imbalance in unit costs and an increased need for grant funding to make up the conservation deficit.

A review of the policies relating to the historic environment in England was initiated in the year 2000 to a fanfare of publicity, consultation exercises and staged responses from government. The practical outcomes of this exercise will not be manifest for some time.

The review anticipates considerable benefits in simplifying and corresponding the established national designations and streamlining the consent regime. They make no provision however for introducing measures that will either secure a coordinated approach to neighbourhood conservation areas or, more significantly, a coherent, holistic approach to historic cities.

Additionally, although the embodied material and energy resource value contained within existing buildings is recognised, attention is focused upon that part of the built inheritance which is assessed as culturally significant, and this within a context in which the established conservation mantra of minimum intervention has been superseded by the facilitation and management of change, and where conservation is seeking to position and reconcile itself with contemporary design in architecture and financial interests within the development and property markets.

All told, the direction of conservation policy in England suggests more of a divergence from the sustainability agenda than a convergence.

Chapter 6
Sustainable Cities and Urban Initiatives

Sustainable cities

Concept and key issues

Just as *heritage*, *restoration*, *authenticity* and *sustainable development* mean many things to many people, so too may the concept of the *sustainable city* be elusive to a single definition. There is however more consensus at the theoretical level, if not about the route to realising it in practice. Indeed, the concept is not new – just the name. And like the many different versions of the ideal city of the Italian Renaissance, its predecessors and successors, it is not a fixed objective in place or time but a constantly changing one as societies' expectations and technologies change.

Under the label of the *sustainable city*, the concept has gathered increasing momentum since the publication of the *Brundtland Report* in 1987. The question that it seeks to address is a simple one. How can urban development meet human needs and at the same time ensure ecological sustainability?

A useful working definition is to be found in *Sustainable Cities*, first published in 1994. It reads as follows:

> A sustainable city is one in which its people and businesses continuously endeavour to improve their natural, built and cultural environments at neighbourhood and regional levels, whilst working in ways which always support the goal of global sustainable development.

This definition resonates closely with Sir Patrick Geddes' exposition of the need to promote a balanced relationship between a city and its region, and between the collectivity of cities and the world's finite resources. Additionally, one of the strongest themes in today's advocacy of the sustainable city is that it is an achievable objective only on condition that we view the city as a dynamic and complex ecosystem and manage it as such.

Echoing Geddes further, *Sustainable Cities* concludes with an eight-part 'manifesto', which includes the following: 'The sustainable city will seek to conserve, enhance and promote its assets in terms of natural, built and cultural environments'.

The key issues that contribute to the sustainable city debate at the global scale include relationships to the use of land; the availability and quality of fresh water; the consumption of non-renewable raw material and energy resources; airborne pollution and its effects on health; the origination and disposal of waste; and the quality of urban environments – including socio-economically and their degree of adaptability to future needs.

Research studies suggest that the land issue is fast becoming critical. In a world with a rapidly increasing population and an even more rapidly increasing urban component of it, the relationship between the land consumed by development and the land that remains for cultivation in order to feed that population is just one of many factors to take into account.

In 1900, with a world population well below two billion, only fifteen per cent of people lived in cities. In 2000, with a world population that had just reached six billion, the percentage had risen to fifty. The average for Europe is eighty per cent, which varies from fifty in a country such as Romania to ninety in the United Kingdom.

Studies indicate that cities occupy just two per cent of the world's land surface, account for some seventy-five per cent of its annual consumption of natural resources, and discharge similar amounts of waste.

In an attempt to define an overview of a city's environmental impact, the favoured indicator is that of *ecological footprint*. This measure of the area of land required in order to sustain the activities of a city includes that for the cultivation of food, fibre and wood; that for the mining of resources; and that for the absorption of wastes. Energy is accounted for by calculating the land required to absorb as biomass the carbon dioxide that is produced.

Following this definition, London's ecological footprint has been calculated to be 125 times its actual surface area; Britain's overall ecological footprint, eight times its total surface area.

Agricultural policy in the European Union encourages farmers to set aside land from arable production. This may lead some to consider that there is a surplus of greenfield land on which to build. As a substantial net importer of food and timber and in a world where oil, the predominant energy source for the movement of goods to and within the United Kingdom, is non-renewable, it is unclear how long we can rely on set-aside as a gauge of future needs for land – which, itself, is a non-renewable resource (Figure 6.1).

As with all statistics, the detail of the methodology and the outcome may be questioned, but what is important here is the principle and the conclusion that comes from it. We live in the age of the city and in the age of sustainability. Cities are the focus of consumption and degradation in the natural environment. To achieve a sustainable world we must start with the city.

Figure 6.1 South-East England. Land is a non-renewable productive and leisure resource. Current European Union set-aside policy for arable land should not be relied upon as a gauge of future land needs.

The vision

The notional consensus on the overall concept of the sustainable city and the key issues that concern it are taken forward to consider how they relate to the size, shape, density, layout and distribution of activities in a city, as well as the degree to which the negative aspects of resource consumption and pollution can be mitigated and the positive ones of environmental quality enhanced and others subsumed – including cultural and biological diversity, and intra- and inter-generational equity.

The widely held starting point is that sustainable cities are compact, high-density (relative to established United Kingdom practice) and mixed-use. They are places where the need for daily travel is reduced; walking and cycling are prioritised; public transport is efficient and viable; energy consumption, the emission of pollutants, and the production of wastes are substantially lowered; and economy in the use of land is assisted by the need for less roads. Also, they are well connected to their localities and to each other by public transport. As such, neither size nor shape is a key issue: the imperatives are proximity and accessibility.

With the exception of all post-industrial mechanisms of transport, whether for goods or people, what is being described here bears singular resemblance to a pre-industrial town or city – a model that was almost invariably in a sustainable relationship to its locality.

The exception, of course, is highly significant. Historic cities were reshaped following the first Industrial Revolution: initially, with the arrival of the railways; subsequently, to accommodate the motor car (Figure 6.2). Today, in a global economy, cities have changed from being places of civilisation to conduits of mobilisation: of people, natural resources, products, and wastes.

The sustainable city concept for the twenty-first century depends on a vision that progressively recovers key aspects of the self-sufficiency of the historical model without retreating into it, and at the same time embraces the global dimension of a hinterland that was previously largely local.

An essential component of that vision is the avoidance of twentieth-century iconoclastic approaches to existing cities – whether interpreted through Le Corbusier's projects and writings, programmes of central area redevelopment and slum clearance, the *Buchanan Report*, or other – and the recognition that much of the physical fabric of existing cities itself constitutes an enormously diverse and rich non-renewable environmental resource, one that is inseparable from the equally rich diversity of the socio-economic frameworks that it harbours. It represents an environmental capital that we have inherited, can care for and adapt creatively, and hand on in a far better condition than

Figure 6.2 Budapest, Hungary: Nyugati railway station. City centre railway stations are major transport interchanges for: inter-city, suburban and underground railways; buses and taxis; bicycles and pedestrians. The *sustainable city* concept depends upon well-integrated and connected cities.

Sustainability and energy-efficient construction

Olympic Village, Homebush Bay, Sydney, Australia

Greenfield developments can serve as test beds for organisational and technological solutions that are directly relevant to historic cities.

The Sydney 2000 Olympic Games were known as the 'Green Olympics' by virtue of the Greenpeace award-winning design of the Olympic Village. Photovoltaic cells and wind power were installed to supply its electricity, solar panels to heat its water, and the village is served by water and waste recycling systems (Figure 6.3).

Figure 6.3 Sydney 2000 Olympic Village. View from the main stadium showing part of the solar cell installation that generates electricity for the development. The Olympic Village was planned so that all of its electricity needs would be met by renewable sources.

The village is laid out for pedestrian and cycle priority, and connected by rail to the city region (Figure 6.4). The buildings were designed to be energy efficient, and no rainforest timbers or polyvinyl chloride (PVC) were used in their construction.

As a material, PVC has been subject to criticism on several counts – including for the amount of energy that is consumed in product manufacture, its shorter lifespan compared to traditional alternatives, and difficulties in recycling.

Over-arching goals

The goals for reduced metabolic flows in cities include:

- hundred per cent renewable-based electricity and heating;
- eighty per cent commuting by non-automobile means;

Figure 6.4 Sydney 2000 Olympic Village. The concourse-level railway station (to the right of the photograph) leads to the platforms below and connects to the city region.

- hundred per cent recycling of solid wastes;
- sewage used for energy extraction and nutrients for farm soil;
- rainwater used locally;
- no PVC or non-recyclable materials used; and
- no rainforest timbers used.

is the case now. In short, it comprises *heritage* within the broad United Nations Educational, Scientific and Cultural Organisation (UNESCO) definition.

The goal of sustainability in a city is the reduction of its use of non-renewable natural resources and production of wastes whilst simultaneously improving its liveability. The implementation of that vision is recognised to be largely a matter of cross-sector coordination and the creative use of modern technologies. Together, this encompasses all of the materials, products and energy consumed in the construction and use of buildings, as well as the '3 Rs' of non-renewable resource and waste management: reduce, recycle and reuse.

For the architect-planner in historic cities, the point of departure is the established infrastructure and buildings: from a sustainability point of view, irrespective of their architectural and historic interest; and from a conservation perspective, as a major added reason for their retention and proper care.

Urban villages

In the introduction to his book *A Vision of Britain: A Personal View of Architecture*, published in 1989, the Prince of Wales wrote: 'Through an organisation called Business in the Community . . . I am hoping we can encourage the development of *urban villages* in order to reintroduce human scale, intimacy and a vibrant street life. These factors can help to restore to people their sense of belonging and pride in their own particular surroundings'.

Three years later this vision was set out by the Urban Villages Group in its report *Urban Villages: A Concept for Creating Mixed-Use Developments on a Sustainable Scale*. The report examined small market towns and mixed-use neighbourhoods in cities – including Edinburgh, London and Paris – and worked to a thesis that applauded their community spirit, social mix, diverse and flourishing micro-scale economies, and adequately healthy property markets (Figures 6.5 and 6.6). It contrasted these with those parts of cities where functional separation and environmental degradation had led to a breakdown of a sense of pride, a growth of disaffection and social unrest.

The report saw the urban village as a model for new development and leant heavily on the theoretical work of the Luxembourg-born architect-planner Leon Krier. Krier was already working for the Prince on plans for the development of Poundbury on the outskirts of Dorchester, a development that was seen as a prototype for a compact, mixed-use, urban village with general pedestrian primacy; in which – theoretically, at least – people can live, work, shop and enjoy an active social life within a single area. Poundbury was also a test bed for the Prince's ideas on the use of traditional building materials and architectural styles (Figure 6.7). The concept relates closely to the New Urbanism movement in the United States.

The imaginary urban village shown in the report is called Greenville, which recalls Ebenezer Howard's vision for a Garden City. Greenville lacked functional separation, the cornerstone of the Garden City concept, but it conjured up a similar romantic image of unitary family life 'in England's green and pleasant land'. As with Howard's vision, the planned unit was to be limited in size: Greenville for a population of up to 5,000 compared to 30,000 in the Garden City; similarly, larger-scale development was to be to a polycentric model. The pastoral image was underscored by the artless new planning designation that was proposed for an urban village: a Structured Planned Urban Development, with the stated acronym of SPUD – the colloquial term for a potato.

The urban village was an excellent concept that lacked both ambition and relevance. Starting as it did from the model of how existing communities in traditional urban situations work, and from the premiss that the reordering of many, especially larger, cities in modern times in Britain had led to a loss of the qualities it sought to recover, the most obvious application of the urban village idea was not to green or brownfield sites but in the developed areas of existing cities, cities which historically across Britain and continental Europe are defined by their mixed-use neighbourhoods.

118 Conservation and Sustainability in Historic Cities

Rothenburg-ob-der-Tauber: *view from across the river Tauber.*

Rothenburg-ob-der-Tauber: *Markusturn* (St Mark's tower), one of the city's landmarks.

Figures 6.5 and 6.6
Rothenburg-ob-der-Tauber, Germany. One of the models for an *urban village* is the medieval market town, including for: its self-sufficiency for daily needs; its compactness and the proximity by foot of its many activities; the quality and sense of permanence of its physical environment its focal points and the legibility of its street pattern.

Figure 6.7 Bratislava, Slovakia. The Urban Villages report recommends public art for its role as a familiar point of reference, in enhancing a sense of place, and as a talking point that brings people together. Here, the statue of Napoleon nonchalantly observing the market square from behind a bench recalls his visit to the city in 1809, four years after the Treaty of Pressburg (the former name for Bratislava) was signed there following the battle of Austerlitz.

Significant parts of major cities such as London and Paris still function in many everyday respects as a series of urban villages.

The concept has potential relevance not merely in the recovery of community pride in the inner parts of historic cities that have lost it or where it has been undermined, but in establishing pride within newer areas – especially post-Second World War housing estates – which lack it. It is the inner cities and peripheral estates where disaffection and socio-economic tensions were highest when the Urban Villages report was published and where they remain highest today.

With such a lack of perceived relevance, the idea of reintroducing traditional urbanity became so watered down that any new development that incorporated more than one use, even at the scale of a part of a city block where houses were built alongside offices or shops, was marketed as an urban village. As a concept for re-establishing – rather than seeking to create – 'mixed-use developments on a sustainable scale', urban villages remains an extremely valuable one. Its application as a key component of sustainable development within existing communities and in historic cities may be considered long overdue.

The ideal sustainable city?

Since ancient times, architects and town planners have sought to define parameters for the ideal city in terms of its size and shape.

Sir Ebenezer Howard's concept for the Garden City, published in 1898, was based on a circle enclosing 400 hectares and housing a population of 30,000 (namely, at a density of 75 persons per hectare). The diameter of the circle was 2.3 kilometres.

The Urban Villages Group's concept for Greenville, published in 1992, was for a compact site of 40 hectares housing a population of 3,000–5,000 (namely, at a density of 75–125 persons per hectare). One of the determining factors of size was the ability to walk at leisure between the furthest points within ten minutes (across a distance of 900–1200 metres depending on shape).

Present-day theory relating to the *sustainable city* suggests that there is neither an ideal size or shape.

Size

The analogy with an ecosystem has been used to suggest that there is no optimal overall size in terms of population. Rather, that a mature sustainable city – like a mature ecosystem – is

Figure 6.8 Vallingby, Stockholm, Sweden: the town square. It has long been the policy in Stockholm to focus dense, mixed-use development around the city's suburban railway system. Vallingby, developed in the 1950s and with a planned population of 23,000, was the first of these satellite centres. In spatial planning terms the organisation of these sub-centres has been likened to stringing together the pearls on a necklace. In the 1980s, contrary to international trends, Stockholm witnessed a decline in car use per capita and an increase in the number of trips by public transport.

dense and compact, highly efficient in the use of space, and has the potential to create increasing efficiencies proportionate to its size. These include in the use of energy and the viability of recycling systems for nutrients and materials.

Additionally, a high level of *functional diversity* improves the balance between producers, manufacturers and services; of *structural diversity*, the availability and proximity of adaptable accommodation to meet changing functional and spatial needs; and of *social diversity*, the balanced and self-regulating community structure that offers overall stability to the system.

This is the theoretical potential based on the ecosystem analogy. To function effectively, it is broken down in the natural world into a complex web of self-contained and balanced species systems, and in the human world into self-supporting and overlapping socio-economic structures.

Shape

In a nation such as the United Kingdom, with an enormous inherited wealth of urban infrastructure and buildings, we do not start with a blank sheet of paper on which architects and planners may choose to draw up ideal plans for linear, circular, ring, mono, or polycentric cities of specific or indeterminate overall size – cities that lack established communities. We start with existing cities in all their variety of size, shape and socio-economic structures. The challenge is to assess and reorder them progressively for sustainability, each according to its individual pattern and needs, including their anticipated population stability or expansion (Figure 6.8).

Urban Renaissance

Launched in 1998 under the chairmanship of Lord (Richard) Rogers, the Urban Task Force published its report *Towards an Urban Renaissance* the following year. Its mission statement set out the brief:

> The Urban Task Force will identify causes of urban decline in England and recommend practical solutions to bring people back into our cities, towns and urban neighbourhoods. It will establish a new vision for urban regeneration founded on the principles of design excellence, social well-being and environmental responsibility within a viable economic and legislative framework.

The Urban Task Force was operating against a backdrop of widespread and continuous decline in the populations of inner cities and the quality of their environments and rates of unemployment that were double the average elsewhere, as well as within a policy context that projected an overall need for homes for an additional 3.8 million households in the period 1996–2021 and a government target that sixty per cent of these should be built on previously developed – or brownfield – land. Brownfield land includes vacant and underused buildings, and the report identified that around 1.3 million residential and commercial properties were currently empty.

The report was focused as much on quality as quantity: above all, on influencing established negative attitudes in England towards cities and presenting them once more as attractive places in which to live, work, and as focal points for social activity. The emphasis on quality first, quantity after, was echoed in the foreword to the report, written by Pasqual Maragall, elected mayor of Barcelona in 1982, in which he described this emphasis as the 'trick' in the renaissance of his city, starting well before the 1992 Olympic Games.

Towards an Urban Renaissance recognised that the concept of a renaissance depended first and foremost on a vision based on participation and shared commitment, and at the dawn of the new millennium the report identified three timely driving forces: the technical revolution centred on information technology and networks of contact and exchange; the ecological threat and greater understanding of the importance of sustainable development; and the social transformation arising from increased life expectancy and a broadening choice of lifestyles.

Towards an Urban Renaissance contained a total of 105 recommendations covering design, transport, management, regeneration, skills, planning, and investment.

The report set out a vision for compact well-designed cities, with close proximity between where people live, work and spend their leisure time, well integrated by public transport, and adaptable to change. These were identified as the principal ingredients for a renaissance in urban environments. They marked a significant shift away from the established mainstream of town planning and towards the concept of the sustainable city.

Diagrammatic examples were given of the principles of mixed-use neighbourhoods and the spatial planning structures that offer well-connected urban networks. If not a comprehensive blueprint for sustainable urban development, *Towards an Urban Renaissance* presented key components in a coherent way and offered a framework within which individual regeneration initiatives, from the scale of single buildings to those promoted through urban regeneration companies, could be accommodated.

The Urban Task Force's report was related to other government initiatives of the time on social inclusion, including the New Deal for Communities aimed at transforming the most deprived estates, and major funding programmes for social housing and regeneration. Compact, well-managed urban neighbourhoods were noted in the report as essential to support local services and foster a sense of community belonging, and although issues relating to education, health, welfare and security fell outside the remit, their importance was adduced.

Whereas *Towards an Urban Renaissance* admitted that 'since the industrial revolution we have lost ownership of [and confidence in] our cities' and that they had been undermined by economic dispersal and social polarisation, the report placed heavy emphasis on the proposition that to be successful, regeneration has to be design led. That this should be so is hardly surprising given Lord Rogers credentials as a world-renowned design architect (Figure 6.9).

Figure 6.9 Paris, France. Richard Rogers achieved international recognition with the completion in 1977 of the Georges-Pompidou Centre, place Beaubourg – immediately to the west of the Marais quarter (Richard Rogers and Renzo Piano, architects). Rogers' interest in cities was first highlighted in his 1995 BBC radio Reith Lectures, which were reworked and published as *Cites for a Small Planet* in 1997. In this, he sets out seven facets of the sustainable city, namely, that it is just, beautiful, creative, ecological, a place of easy contact, compact and polycentric, and diverse.

This emphasis was not without its critics, who felt that the insistence on a fashion-led approach to urban design favoured by some professionals risked alienating an important sector of public opinion that was not in tune with it. It also deflected attention from the issue that was more central to a nationwide urban renaissance, namely, a balanced strategy for securing long-term job opportunities in established communities in locations where brownfield sites were concentrated. The urban renaissance and sustainable communities agenda would then be working in harmony.

The report did, however, stress that quality of design is not simply about creating new developments but about making best use of existing urban environments, from historic central areas to low-density suburbs, repairing the existing urban fabric, and recycling underused buildings. An important recommendation was that empty property strategies should be established in every local authority area; another, that public bodies should release redundant urban land and buildings for regeneration (Figure 6.10). The report also stressed the importance of tailoring general policies and programmes to each and every individual location.

Over the issue of movement of people, *Towards an Urban Renaissance* recognised that one of the best ways to encourage and absorb more people into urban areas is to reduce the need to travel by car, coupled with policies that discriminate in favour of walking, cycling and public transport. The report

Figure 6.10 Derby, England. The former Midlands Counties and North Midland Railway Company complex of locomotive and carriage workshops, including a roundhouse, sheds and smithies, dates from 1839. Its principal buildings are largely intact, but boarded up and in poor condition. Owned by Derby City Council and situated on land that adjoins the city's railway station, it offers a major opportunity for a mixed-use regeneration scheme on a brownfield site.

supported the introduction at neighbourhood level of 'home zones' – the mechanism for establishing pedestrian priority in the public realm. The Urban Task Force also made the important connection between place and people: 'cities make citizens, and citizens make cities'.

Lord Rogers' follow-up report *Towards a Strong Urban Renaissance*, published six years later in 2005, noted a measurable change in attitude in favour of cities, reflected not least by substantial population increases in the central areas of post-industrial cities such as Manchester and Liverpool, as well as significant increases in investment by the private-sector and in urban public transport infrastructure. Furthermore, it noted that a national average of seventy per cent – not sixty per cent as targeted by the government – of new development was now taking place on brownfield land.

On the downside, this follow-up report noted increased inequalities in cities, reflected not least by the inexorable increase in house prices in many parts of the country and the difficulties this posed for households on low incomes. It also reported a lack of focus on design quality, integrated planning, and sustainable urban development.

Over the six years between the two reports the probabilities of a balanced approach to an urban renaissance were upset by two complementary government actions:

First, the encouragement given to the relentless expansion of job opportunities and housing demand in the South-East of England coupled with attempts to create sustainable communities outside city centres.

Copenhagen: cities make citizens, and citizens make cities

The city centre

Experience in Copenhagen suggests that a focus on people in the public realm – including ease and safety of movement, air quality, and related urban management – is at least as important as a design-led approach where urban renaissance is concerned.

By the early-1960s Copenhagen was evolving as a typical Western city on the Anglo-American model, with separated land uses, low-density suburban expansion, and increasing use of the private car to travel to work in the city centre (Figure 6.11).

Figure 6.11 Copenhagen, Denmark. Nyhavn was formerly a busy part of Copenhagen's harbour and full of seamen's bars. By the 1960s it had become a commuter parking lot – as can be seen in this 1968 photograph. Cleared of cars in 1980 and turned into a waterside pedestrian area, it is now one of the most popular places to eat, drink, meet, and just sit.

The city inaugurated a long-term incremental programme for the historic centre directed at reducing the number of parking spaces for commuters, pedestrianising the streets and squares, and creating low-speed zones.

The historic centre has a medieval street pattern. In 1962 its main shopping street, Strøget, was the first to be pedestrianised. Progressively, inner city housing was built or refurbished, the public realm relandscaped, sculptures and seating introduced, and street buskers, entertainers, markets, festivals and other street-based activities supported. As the architect Jan Gehl has written: 'We decided to make the public realm so attractive it would drag people back into the streets, whilst making it simultaneously difficult to go there by car' and 'The city became like a good party'.

Investment and people have been encouraged into the city centre, and social and recreational activity has tripled in the city's main streets. There is no congestion charging, and parking charges in the central area are high in order to encourage short stay, high turnover of spaces. There are no underground shopping malls and no high level walkways. Everything happens at street level.

The revival of confidence in the city centre has been achieved through a focus on people, public spaces and public life. The design philosophy is straightforward: simplicity and elegance to harmonise with the historic environment, the use of natural materials, the low-key insinuation of contemporary and experimental elements, and strict control over advertising and signage.

The city region

Copenhagen has changed from a car-orientated into a human-orientated city. City-wide, there is a culture of respect and priority for pedestrians and cyclists, and a third of all commuter journeys are made by bicycle.

At the sub-regional scale, the city plan has extended along the radial railway routes, establishing corridors of growth and mixed-use urban villages focused on the stations. A declining demand for single, detached houses on the urban fringe has been reported: their residents feel isolated from the life of the city.

The follow-up phase in the development of the city has been the introduction of a light railway system concentrically between sub-centres, paid for out of the land development opportunities that it is creating.

Second, the Housing Market Renewal Initiative, better known as Pathfinder, that focused on nine areas in the North of England. Involving a public investment of 1.2 billion pounds over the period 2003–08, this was directed at recovering values in the housing property market by large-scale programmes of demolition and new building. Likened to a latter-day programme of slum clearance, this was initially estimated to involve the demolition of 400,000 predominantly terraced houses – of the kind that obtain premium prices in the South-East.

Despite Pathfinder's concentration in parts of the country that have lost their traditional industries, there is reported to be an overall shortage across the region of low cost housing to accommodate the increased number of households. Reducing the supply of houses and thereby driving up prices will marginalise households on low income still further. Save Britain's Heritage, in its publication *Pathfinder*, quotes Albert Einstein: 'The definition of insanity is doing the same thing over and over again, and expecting different results'.

However many houses end up being demolished, each and every one represents environmental capital. Equally relevant even at the basic level of water supply, the South-East is increasingly prone to annual shortages, whereas in the North of England there is an abundance. There is a lack of strategic thinking here.

To demonstrate its commitment to environmental responsibility – safeguarding greenfield land from development, avoiding further erosion of the countryside and the depletion of energy resources – *Towards a Strong Urban Renaissance* proposed raising the minimum density standard for new residential development to 40 dwellings per hectare.

This proposal evoked a dissenting footnote from one of the Task Force members, Sir Peter Hall. In this, he rejects the need to save greenfield land on the premiss: first, that 'we have a surplus in South-East England'; second, that 'present policies are already ... causing an unprecedented increase in apartment construction, unsuitable for families with children and undesired by potential residents'. Debatable as both of these propositions may be, the Pathfinder initiative involves – as we have seen – the demolition of many thousands of family homes in the North of England. Furthermore, reports suggest that some at least of the new apartments being built in the South-East have been constructed to lower space standards than in the poorest countries of Eastern Europe.

Empty properties in historic cities

Lord Rogers concluded his introduction to *Towards an Urban Renaissance* with the simple statement: 'An urban renaissance is desirable, necessary, achievable and long overdue'. The starting point is the existing urban environment, and the Urban Task Force's report emphasises the importance of making best use of it.

Towards an Urban Renaissance quoted the conclusion of a survey conducted in 1998 to the effect that there were 735,000 empty dwellings across England. Additionally, it has been estimated that more than 300,000 new homes could be created by converting the vacant upper floors of commercial properties in city centres (Figure 6.12). The figure of 73,000 is given for London alone.

The Urban Task Force recognised that this all represents not merely the opportunity to satisfy a substantial part of the additional housing need, but that a mix of residents in the commercial hearts of cities creates the critical mass that is vital to economic and social regeneration. Further, this enhances security in city centres: of property and in the public realm.

A series of coordinated initiatives to bring upper floors back into use stems from the Living Over The Shop project (LOTS) set up by the Joseph Rowntree Foundation in 1989. Although some successful projects had been implemented in individual towns and cities before the publication of LOTS' first report a year later, and some momentum gathered through the 1990s, a series of factors continues to militate against a significant breakthrough.

These include owner resistance, legal issues over mixed tenure, physical access, and local plans that adopt a two-dimensional approach to land-use zoning, not a three-dimensional approach to building uses.

Test cases had, however, demonstrated a significant latent and unsatisfied demand – contrary to the assumption that English people are overwhelmingly averse to living in flats, especially in city centres. They also demonstrated that, notwithstanding the historical reality that the majority of upper floors were built to serve as dwellings for the shopkeepers below, problems of access were rarely insurmountable.

By way of comparison, Scottish cities follow the continental European model. Over sixty per cent of the populations of the three largest cities live

Figure 6.12 Derby, England: the market square. It is estimated that homes for over half-a-million people could be created by converting the vacant upper floors of shops in England. Vacant units such as these would be prime residential properties in cities across continental Europe.

in flats, many of which are located in their centres. A vertical mix of uses is the norm, the buildings were constructed with separate accesses, the legal basis of property tenure is favourable, and local plans are three-dimensional. Moreover, many English people live in flats in the centre of Scottish cities, people whose lifestyles do not fit comfortably with the responsibility of a house and garden – with or without children.

The Urban White Paper, published in 2000 in the aftermath of the Urban Task Force's report, did result in fiscal incentives for the conversion of properties to residential use. Little progress, however, has been made over the issue of vacant upper floors in cities. Additionally, a statutory duty on local authorities to produce empty property strategies has not been introduced.

Urban conservation: a metropolitan centre

Paris, France

As related in Chapters 1 and 3, a strategic approach to decentralisation at national and city-regional level, allied with favourable spatial planning conditions at a more local level, established a stable environmental and economic basis for conservation and regeneration across the city of Paris.

Following the initial museological phase in the 1960s and early-1970s, the change of direction in the Marais quarter was significant. With its physical environment no longer destined to be fossilised in a time warp, and its buildings no longer restricted to a limited number of institutional uses, it rapidly became one of the liveliest parts of the city: a large city centre neighbourhood with which people at home and abroad identified enthusiastically. New theatres complemented the established mix of museums and art galleries, an annual arts festival and other performance events took off, and the streets and squares became filled with people seven days a week.

Historic buildings have been restored and empty ones brought back into use, contemporary architecture now complements the nine-century tradition of building, and public spaces have been discretely relaid and replanted (Figures 6.13 and 6.14).

Certainly, the socio-economic profile of the area has evolved from its early-twentieth-century nadir, and the Marais has become a fashionable quarter for artists and intellectuals. Some local shops have changed to become boutiques, restaurants and private galleries (Figure 6.15). However, with a quoted population density of 630 persons per hectare in 1970 and 600 in 1990, the strength of demand created by the overwhelmingly residential character of the area is such that there is a thriving range of everyday and specialist shops; and traditional crafts such as cobblers, tailors, wood-turners and bookbinders can all be seen at work through their ground-floor windows. The contrast with the limited range of multinational, chain and franchise stores that characterise and monotonise historic city centres across the United Kingdom could not be greater.

Figure 6.13 Marais quarter, Paris, France. The ground floor arcade of shops and cafés and the upper floor houses and flats in the Place des Vosges have progressively been restored since the 1960s. Additionally, the central square has been relaid and replanted.

Figure 6.14 Marais quarter, Paris, France. Contemporary infill at the corner of the Rue Vieille-du-Temple and the Rue de Francs Bourgeios.

Figure 6.15 Marais quarter, Paris, France. Some everyday shops such as this *boulangerie-pâtisserie* on the corner of the Rue des Francs Bourgeios and the Rue de Sevigne have changed to boutiques. Their historic shop fronts have frequently been retained.

Ile-Saint-Louis: *bakery*. Ile-Saint-Louis: *joiner and cabinetmaker*.

Figures 6.16 and 6.17   Ile-Saint-Louis, Paris, France. Even in the most central locations of the French capital there is an enormous range of small shops and workshops serving everyday and specialist needs. The density of the residential population supports this diversity.

It was Baron Haussmann (1809–91), working under Napoleon III, who introduced the planning laws that protect small shopkeepers and other traders in Paris, not least by placing a presumption against the amalgamation of ground-floor commercial units. Ironically, it was the latter's uncle, Napoleon I, who famously and disparagingly referred to the English as 'a nation of shopkeepers'. The Marais is typical of other quarters of Paris (Figures 6.16 and 6.17), and Paris is typical of France.

As with New Yorkers, one of the characteristics of Parisians is the comparatively high proportion that do not own cars. They argue that they have no need: they have proximity and an excellent, integrated, public transport system. On the occasions when they need a car, they hire one: it is more practical and cost effective.

Chapter 6: digest

The sustainable cities concept seeks to balance human needs and the aspirations of citizens at local level with ecological sustainability at the global scale. It recognises that cities are the focus of consumption and degradation in the natural environment, and that to achieve a sustainable world we must start with the city.

Issues at the forefront of the concept include the use of land, the supply of water, the consumption of non-renewable material and energy resources, the reduction of pollution, the containment and recycling of wastes, and the quality and adaptability of urban environments.

The sustainable city seeks to conserve and enhance what exists in the natural, built and cultural environments. It views the city as a dynamic and complex ecosystem, one in which a core objective is the achievement of a balanced and self-regulating socio-economic and environmental organisation based on functional, structural and social diversity.

Sustainable cities are expressed as compact, dense, mixed-use, and economical in the use of land; places where proximity and pedestrian movement are prioritised, and public transport is efficient and well integrated.

The vision seeks to recover key aspects of the self-sufficiency of the pre-industrial city without retreating into it, embracing the global dimension of a hinterland that was once local. It recognises that implementation is largely a matter of cross-sectoral coordination and the creative use of technologies. From the conservation perspective, the vision recognises that the environmental capital of existing buildings and infrastructures is at least as important as their cultural value.

In recent years, a number of initiatives have contributed positively to the sustainable cities debate.

Urban Villages explored the idea of creating mixed-use developments at a small, neighbourhood scale, and has wider relevance to existing cities and their communities than was represented by its authors.

The *Urban Renaissance* took a much broader view and explored key components of an over-arching framework for sustainable urban development across the length and breadth of England. It sought to inspire a revival of confidence in cities and citizenship and to match this to increasing public awareness of the sustainability agenda and changes in society. As such, the report may have been better advised to place environmental responsibility and social well-being above urban design in its list of priorities. That citizens are more important to cities than design is attested by experience elsewhere.

Much of the wisdom in the *Urban Renaissance* has been undermined by a lack of strategic planning at national level, including in the exacerbation of the North–South divide and a failure to address proactively the opportunities afforded by the enormous range and quantity of empty and underused property in English cities. Their environmental capital is one of the keys to unlocking the untapped potential of urban conservation to contribute to sustainable development.

Chapter 7
Managing World Heritage Cities: United Kingdom

International guidance

Provisions under the World Heritage Convention

The *World Heritage Convention* was adopted in 1972 and the first tranche of World Heritage Sites was inscribed in 1978. It was not until 1997 however that the submission of a management plan became a pre-requisite for inscription. Since then, state parties have been called upon by the United Nations Educational, Scientific and Cultural Organisation (UNESCO) to submit periodic reports setting out the legislative and administrative provisions and other actions that have been taken for implementation of the *Convention*, including state of conservation reports for each individual World Heritage Site.

This process anticipates the preparation and submission to the World Heritage Centre of management plans for those sites that were inscribed before 1997. Of the two United Kingdom historic cities on the World Heritage List, the city of Bath was inscribed in 1987 and its management plan published in 2003, and the Old and New Towns of Edinburgh were inscribed as a site in 1995 and the management plan published in 2005.

The process also anticipates that state parties will review the need for revisions to the legislative and administrative frameworks within which management is exercised.

Recommendations and Charters

The 1972 *Recommendation Concerning the Protection, at National Level, of the Cultural and Natural Heritage* was effectively the first international document to amplify the relationship between cultural heritage, communities, socio-economic development and the historic areas of cities (see page 66). In 1975, the Council of Europe expanded upon this thesis with its promotion of the concept of integrated conservation (see pages 12 to 13).

UNESCO adopted the *Recommendation Concerning the Safeguarding and Contemporary Role of Historic Areas* at its General Conference in Nairobi in 1976, urging member states to apply it at national level.

This 1976 *Recommendation* asserted the societal importance of historic areas: their role in defining cultural diversity and the identity of individual communities, and the need to integrate them into 'the life of contemporary society [as] a basic factor in town-planning and land development'. This *Recommendation* noted the frequent absence at national level of legislative provisions that related the architectural heritage to its planning context, reproached the social disturbance and economic loss resulting from the destruction of historic and traditional areas, and urged 'comprehensive and energetic policies' for their protection, renovation and revitalisation, integral with their surroundings.

Importance is attached in this document to continuity of human activities in historic areas – however modest they may be, including traditional living patterns and crafts – on an equal footing with protection of the buildings, established plot sizes, street patterns and overall spatial organisation. Awareness raising within communities together with continuity in the training of skills at all levels are also emphasised.

The 1987 *Washington Charter* elaborated this 1976 *Recommendation* within the non-governmental context of the International Council on Monuments and Sites (ICOMOS) (see pages 13 to 14).

The UNESCO Operational Guidelines

The UNESCO *Operational Guidelines for the Implementation of the World Heritage Convention* were most recently revised in February 2005. They clarify and expand upon a number of key concepts and issues.

Authenticity and integrity

These are referred to in Chapter 4.

Protection and management

The *Operational Guidelines* require that sites have adequate protection and management mechanisms in place, and that the conditions of authenticity and integrity that existed at the time of inscription as a World Heritage Site will be maintained or enhanced in the future; also, that the boundaries of a site are drawn widely enough to enable this to be achieved – both directly and by association. Additionally, there is a presumption that sites should have an adequate buffer zone, the purpose of which is to protect the setting of the site.

The *Operational Guidelines* set out general expectations for a management plan – for example, that it should include provisions for shared understanding and the involvement of stakeholders, capacity building, and the allocation of necessary resources – but offer no detailed guidance as to scope and content.

The ICCROM Management Guidelines

The International Centre for Conservation in Rome (ICCROM) *Management Guidelines for World Cultural Heritage Sites* are written very much from the perspective of the architectural conservator and are of most relevance to sites that are under close curatorial control.

The chapter on historic areas does however provide general commentary on the qualities of historic areas and the threats posed by development pressures, and sets out some key planning objectives, of which the over-arching one is control of the rate of change to fabric and community alike. This is also stated as minimum intervention.

The same chapter highlights the relationship between sustainable development and the management of resources, and affirms that urban conservation is not simply a question of the architectural framework of a historic area but one related to the human values of social and economic context coupled with the maintenance of appropriate functions and, where feasible, traditional types of use. It goes on to establish the important principle that the closer the present and future uses of the existing buildings in a historic area are to the purposes for which they were built, the lower will be the costs and the better it is for the area as a whole.

Successful urban conservation is acknowledged to require the involvement of many different professionals, including city planners, architects, sociologists and administrators. The key issues identified in the *Management Guidelines* can be summarised as:

- the need to treat a historic centre in the context of the wider city;
- the need to adapt standardised planning techniques to suit local conditions, historic urban texture and scale, adopting a bottom-up rather than a top-down approach;
- the need to respect the intangible cultural traditions of a historic city;
- the importance of simple buildings and vernacular architecture in distinguishing a historic city from a group of monuments;
- the prevention of out-of-scale uses and buildings (including tall buildings);
- the importance of treating the existing historic fabric on equal terms with other factors in the general planning process;
- the principle that environmental capacity should be the determining factor in transport and traffic planning;
- the importance of securing beneficial use within the community through a mixture of residential, commercial, industrial and leisure activities that accord with the scale of the existing buildings and urban grain;
- the need to avoid both facadism and architectural pastiche;

- the limitation of new construction to infill that respects the scale and character of its historic context, for which several pointers are listed, including rhythm, mass, street boundary line, silhouette, traditional or compatible materials, window to wall ratio, quality; and
- the importance of regular maintenance using traditional materials and building techniques.

It is unfortunate that having gone to the trouble to describe these issues that the *Management Guidelines* conclude with a very limited summary of what the management of a historic urban area actually involves, namely, the analysis of urban morphology, (conservative) property management, modest rehabilitation schemes, and social input and consultation with occupants.

Context

There is a strong sense that cultural and natural heritage sites on the World Heritage List should act as exemplars for the management of cultural and natural heritage generally at national and international levels. World Heritage Cities such as Bath and Edinburgh engage with a series of concepts and agendas: cultural landscapes, tangible and intangible heritage, authenticity and integrity, and sustainability at the urban scale. The responsibility and opportunity to integrate these into the management plans for the two United Kingdom sites is there to be seized.

As can be seen, however, although the UNESCO *Operational Guidelines* and the ICCROM *Management Guidelines* provide markers, they do not provide unequivocal guidance as to the range of issues that are expected to be addressed within the long-term management plan for the historic areas of cities. It is not surprising therefore to find a variation of emphasis between the management plans for the city of Bath and the Old and New Towns of Edinburgh World Heritage Sites.

The Bath and Edinburgh World Heritage Sites

The cities

The World Heritage Sites of Bath and Edinburgh are highly distinctive but share many features in common.

Both sites enjoy dramatic landscape settings: Bath is enclosed in a deep, wooded valley on a curve of the river Avon and encircled by hills; Edinburgh is built astride a complex topography and geology in a panoramic setting that commands distant views inland and across the wide estuary of the river Forth. Both sites possess a sense of place and physical identity that depends on the quality, consistency, and aura of permanence and timelessness with which their built environments have been harmonised with their landscape settings (Figures 7.1 and 7.2).

Figure 7.1 Bath, England. The city of Bath was inscribed on the World Heritage List in 1987 in recognition of its importance as a thermal spa in Roman times, its role as a centre for the wool industry in the Middle Ages, and the elegance and harmony of its development as a neoclassical city in the eighteenth century. Lansdown Crescent (built 1789–93; John Palmer, architect) – with its sinuous concave and convex curves that lead the eye along, round and down the contours – gives the impression of being an integral part of the landscape.

Figure 7.2 Edinburgh, Scotland. The Old and New Towns of Edinburgh were inscribed on the World Heritage List in 1995 in recognition of the individual importance and harmonious juxtaposition of the medieval Old Town with the neoclassical late-eighteenth to mid-nineteenth-century New Town, and for the far-reaching influence of the latter on European urban planning. To the north, the Edinburgh site commands views across the estuary of the river Forth. The first New Town was originally developed as a self-contained residential suburb for the aristocracy and the professional and merchant classes; it is now the focus for major retailing activity in the city.

In both cities, visual homogeneity across the architecture of diverse periods relies on the dominant, locally sourced, building material, stone – at Bath, honey-coloured limestone; at Edinburgh, buff-coloured sandstone that weathers to grey – coupled with a limited palette of supporting materials: natural slate at roofs; cast and wrought iron at railings, balconies and street lamps; fine joinery and glazing at doors and windows; and stone at footpath paving, kerbs, and roadway setts.

In both cities, the first evidence of human settlement dates from prehistoric times. Bath's rich history relies on its status as a spa town from the Roman period onwards; Edinburgh's, primarily today, as Scotland's capital since the fifteenth century.

Until the eighteenth century, both cities were constrained within their medieval bounds, when first Bath, then Edinburgh, expanded dramatically. In the case of Bath, the neoclassical city was overlaid on the earlier settlement and extended out from it (Figure 7.3); today, the legacy of town planning and

Figure 7.3 Although the medieval street pattern is still discernable on plan, Bath Abbey is one of the few buildings in the city to have survived from the Middle Ages. The present structure, built over part of the former Norman cathedral, dates from 1499 and is one of the last great churches in England to have been built in the Gothic perpendicular style.

Figure 7.4
Edinburgh, Scotland. The New Town was built in seven phases, and nine-tenths of it was originally constructed as housing. The first phase was laid out in accordance with James Craig's plan of 1767 and Charlotte Square is one of its few formal compositions. Its north side was begun in 1791 to the design of Robert Adam (1728–92). As with the north side of Queen Square, Bath (see Figure 1.4), its palace facade unites a row of terraced houses. The neoclassical architecture of first Bath and then Edinburgh was greatly influenced by the Italian architect Andrea Palladio (1508–80).

architecture from the Georgian period dominates. By contrast, in Edinburgh the neoclassical city was built apart from the medieval city, which in turn was subject to extensive remodelling and improvement from the late-eighteenth century onwards (see pages 31 to 32).

Common to both World Heritage Sites is their place in the history of neo-classical architecture and town planning from the eighteenth into the nineteenth centuries and the progressive influence they had, first on each other and then more widely across Europe (Figure 7.4).

Today, the two cities are regional centres for employment, shopping and leisure; prominent in higher education; international tourist destinations attracting well over three million visitors annually; and, since the late-1940s, both have progressively developed year-round calendars of festivals and events in the performing arts and sports. Edinburgh, additionally, is a financial and administrative centre of national importance, home to Scotland's national museums, galleries and library and, once again, a seat of Parliament (Figure 7.5).

Figure 7.5 The development of the classically planned Edinburgh New Town separate from the medieval Old Town represents an early example of complementary urban development as opposed to destructive redevelopment. In town planning terms it anticipated the teaching and practical work of Gustavo Giovannoni in Italy during the first half of the twentieth century. The former Nor' Loch, now Princes Street Gardens, separates the two parts of the historic city. The skyline of the Old Town is dominated by Edinburgh Castle – a medieval fortress that was extended under the Scottish monarchy in the sixteenth century and again when it became a military barracks in the eighteenth. The Mound – shown here in the middle distance – forms the westernmost connection between the Old and New Towns and houses: to the right, the Royal Scottish Academy building (built 1823–35; William Playfair, architect); to the left, the National Gallery of Scotland (built 1845; William Playfair, architect).

Post-Second World War threats

In the 1940s the two cities were the subject of city-wide studies led by the town planner Sir Patrick Abercrombie (1879–1957), studies that anticipated major reordering and development especially in their historic centres. For Bath, the visual presentations show a tightly encircling ring road, partly within the line of the former medieval walls, and several major projects of redevelopment – including for the bus station – in a formal classical manner reminiscent of the visionary plan prepared by John Wood the Elder in 1725. This had aimed to transform Bath into a monumental classical city in the Roman tradition.

Abercrombie's plan for Edinburgh showed a network of new highways encircling and cutting through the Old Town to form a ring; also, connecting across the whole of the city. Proposals to implement variations of this network resulted in neglect and planning blight along the chosen routes until the mid-1970s, when the road schemes were abandoned and the worst affected streets declared housing action areas for improvement.

Bath had also suffered serious aerial bombardment in 1942, when amongst scores of other buildings the Assembly Rooms and houses in Queen Square, the Circus, and Royal and Lansdown Crescents were destroyed; architecturally important buildings were subsequently rebuilt. The various plans and studies prepared by Sir Colin Buchanan for Bath in the 1960s are referred to in Chapter 2.

From the 1950s through to the early-1970s both cities suffered significant losses to their built environments. In Bath, great swathes of artisan housing were levelled, much of it in the name of slum clearance; in Edinburgh, two squares (George Square partially, St James Square totally), some Georgian tenement housing on the north-western edge of the New Town, and a number of important individual buildings – including on Princes Street – were lost.

Over the same period, also in both cities, national and local amenity societies, academics, practising professionals and sections of the media fought long and hard to redress the balance – not just for the key individual buildings and the monumental set pieces, but for the full range of building types and sizes down to the sides streets, shopping alleys, and tenements built for the serving classes (Figure 7.6). As a result, exemplary programmes of

Figure 7.6 Margaret Buildings, Bath, a pedestrian shopping street close to the grandeur of Royal Crescent. This street was excluded from the 'buildings worthy of preservation' in the 1945 Abercrombie plan.

Bath Preservation Trust

Saving Bath

Articulate and campaigning amenity groups have long been a feature in the United Kingdom. Their skills in channelling and mobilising public opinion have frequently made them highly influential in the cause of urban conservation.

The Bath Preservation Trust was founded in 1934 and is the principal such organisation in the city. The Trust played a leading role in the campaigns of the 1960s and 1970s to *Save Bath*. Today, it is equally vociferous in its defence of the Green Belt that surrounds the city.

Museums

The Trust owns and manages three historic buildings that it has restored as visitor attractions:

- No 1 Royal Crescent (Figure 7.7);
- the Countess of Huntingdon's Gothic style Chapel, commissioned in 1765, which houses the Building of Bath Museum. This tells the story of the development of the Georgian city: how its houses were designed, built and decorated, and how they were lived in. The centre-piece is a scale model of the city; and
- Beckford's Tower and Museum on Lansdown Hill (Figure 7.8).

Figure 7.7 No 1 Royal Crescent, Bath, dates from 1767 and was the first house to be built in the Crescent (John Wood the Younger, architect). It has been carefully restored and furnished with authentic Georgian furniture, paintings, carpets, china and glass, to show how one of the great houses of the period would have been lived in.

Figure 7.8 Bath. Beckford's Italianate tower stands on the northern edge of the World Heritage Site and commands panoramic views of the city and its surrounding countryside. Completed in 1827, it was built for the wealthy collector and patron William Beckford (1760–1844) as his study-retreat.

The Trust is also a trustee of a fourth visitor attraction, the William Herschel House and Museum. This modest, four-storey town house, built around 1764, is famous as the house from which, in 1781, the musician and astronomer William Herschel (1738–1822) discovered the planet Uranus using his own home-built telescope.

Historic buildings grants

Since 1996 the Bath Preservation Trust has been the only organisation in the city to award historic buildings grants. From a very modest budget it supports the replacement of missing architectural features.

historic building conservation were begun and implemented across the central areas of both cities and their historic environments are now in generally good condition. From the perspective of 1970 it was, as they say, a close run thing.

Cockburn Association, Edinburgh

'The unofficial guardian of the city'

The Cockburn Association was founded in 1875 – two years before the Society for the Preservation of Ancient Buildings – and is named after Lord Cockburn (1779–1854), a pioneer activist for the city. Described as 'a popular association for preserving and increasing the attractions of the city and its neighbourhood', it is one of the most broadly active civic societies in the United Kingdom addressing issues as diverse as transport, watercourses, open spaces, the built environment, and the city's surrounding Green Belt.

The Association has been engaged in a variety of campaigns through the planning system and the media, including the following:

Successfully (mid-1950s to early-1970s):

- against proposals for an inner ring road and the construction of multi-storey car parks in Princes Street Gardens;
- in favour of the city's outer bypass; and
- in favour of the retention of Haymarket Station (1840–42, one of the earliest railway stations in Scotland; John Miller, engineer) and against a proposal for an 80-metre-high office block that would have replaced it.

Unsuccessfully (1990s onwards):

- a new rail link between the city centre and the airport.

Additionally, whereas the Association has sought to promote city-wide improvements in traffic management it has not insinuated a balanced strategic approach into the key related issue of land use planning.

Other activities include:

- a building preservation trust that has successfully restored a number of buildings at risk;
- annual *Open Doors Days*, the trans-European initiative during which houses and offices that are not normally open to the public open their doors to visitors; and
- painting competitions for schoolchildren.

The World Heritage Sites

Bath is the only complete city in the United Kingdom on the World Heritage List. The *de facto* extent of the site is defined by the municipal boundary of the former Bath City Council. It extends over 29 square kilometres, contains a population of 84,000, and includes 5,000 listed buildings. A single conservation area covers two thirds of the city, and although there is no buffer zone the setting of the site is protected by designated Green Belt and the Cotswolds Area of Outstanding Natural Beauty. The countryside stretches into the city in several places, creating large tracts of green in the midst of the urban environment.

The Edinburgh site covers 4.5 square kilometres. Its boundary is very tightly drawn within the built-up area and is inconsistent with those of the six individual conservation areas that, to varying extents, overlap it. At 24,000, the population of the World Heritage Site represents five per cent of the city's total. The site excludes significant areas of eighteenth- and nineteenth-century development that adjoin it immediately to the south, east and north; it also excludes key landscape features such as Arthur's Seat within Holyrood Park. These and other omissions were commented upon in the ICOMOS Advisory Report prior to the inscription of the Old and New Towns as a World Heritage Site, with a strong recommendation that they be reconsidered. This has yet to be done; also, there is as yet no defined buffer zone.

The Bath and Edinburgh sites were inscribed under stated criteria that reflect and focus upon their history, architecture and town planning (Figure 7.9). In the United Kingdom such criteria match established government policies and guidance relating to historic buildings, conservation areas, and urban design. The Edinburgh management plan emphasises this by quoting the ICOMOS Advisory Report, which places the site into the single category of a *group of buildings*.

Figure 7.9
Edinburgh. Begun in 1822, the development of the sixth phase of the New Town on the Moray estate came towards the end of the Georgian period. Moray Place, shown here, is polygonal on plan (James Gillespie Graham, architect).

The management plans for the Bath and Edinburgh World Heritage Sites

Authorship

The management plan for the city of Bath was coordinated by Abigail Harrap, World Heritage Officer for Bath and North-East Somerset Council, a unitary local authority and the principal body responsible for managing the site in all its complexity. Guided by an inclusive governmental and non-governmental steering group and supported by a widely representative stakeholder group, the Bath management plan has been conceived and delivered as a partnership document for the city.

The management plan for the Old and New Towns of Edinburgh has been prepared by Edinburgh World Heritage Trust, assisted by a steering group comprising representatives of local and national government only. For all promotional purposes the trust is now known simply as Edinburgh World Heritage. Funded by the City of Edinburgh Council and Historic Scotland, its principal functions are focused on the conservation of historic buildings and aspects of the wider historic environment within the World Heritage Site. It has an advisory role to government, not a statutory one. The management plan lists a limited range of historic environment, local authority and development agency interests as the main parties responsible for managing the site.

Edinburgh: a culture of architectural conservation

Edinburgh New Town Conservation Committee, Edinburgh Old Town Renewal Trust, and Edinburgh World Heritage Trust

Since 1970 urban conservation in Edinburgh has been marked by a sequence of decennial conferences and the related establishment of three key organisations.

The conferences

- 1970: *The Conservation of Georgian Edinburgh*. Informed by a systematic survey that had been coordinated by the Edinburgh Architectural Association in the late-1960s, in what has been described as a 'remarkable exercise in voluntary effort', the purpose of this conference was to state the case for preserving the New Town and to initiate action. This conference focused on the architectural and town planning value of the area and explicitly avoided other environmental, social and economic issues.
- 1980: *The Architectural Heritage: A Maintenance Crisis*. This conference addressed the need for adequate maintenance regimes for historic buildings and coincided with the publication of a comprehensive maintenance manual entitled *The Care and Conservation of Georgian Houses*.
- 1990: *Civilising the City*. This greatly extended the range of issues to embrace strategic planning, public transport, conflicts posed by tourism, and the importance of sustaining community life in historic city centres.

- 2000: *Urban Pride: Living and Working in a World Heritage City.* Extending the range of issues yet further, this conference addressed political, administrative, social and cultural dimensions and their relationship to pride in citizenship.

Edinburgh: *Moray Place.*

Edinburgh: *Heriot Row.*

Figures 7.10 and 7.11 Edinburgh. These examples of ironwork at window balconies and street lanterns are typical of historic features throughout the New Town.

The organisations

- 1970: The founding of the Edinburgh New Town Conservation Committee. Jointly sponsored by central and local government its primary function was as a grant-awarding body for the repair and restoration of historic buildings. It performed a key role in promoting both public awareness and the availability of skills and materials. Also, for several years it funded and managed a business that recycled building components and materials within the New Town, commissioning and marketing replicas of historic features such as street lanterns, and brass fittings for windows and doors (Figures 7.10 and 7.11). The grant-aid policy of this Committee was to focus on the weakest links, tenements in the most run-down parts of the historic New Town – known as the 'tattered fringe' – thereby reviving confidence in the whole of the area. The New Town Committee's work was recognised by the Europa Nostra Medal of Honour in 1987 and Diploma of Merit in 1987–88.
- 1985: The setting-up of the Edinburgh Old Town Renewal Trust. This grew out of the experience of the New Town Committee and was also primarily a grant-awarding body.

However, it was more proactive in the community and in identifying and promoting regeneration opportunities – for which it acted as a catalyst and a coordinator. Its programme sought to extend the concept of architectural conservation into that of sustainable development. One of the many successes in the Old Town has been the recovery of its population: from 20,000 in 1931 it fell to 3,000 in 1961; by 1991 it had recovered to 7,000 and by 2001 to 10,000. During the 1990s the Old Town Trust shared its skills with the Old Town Revitalisation Agency in Vilnius, Lithuania; also, with the community in the Kazimierz quarter of Krakov, Poland.

- 1999: The establishment of the Edinburgh World Heritage Trust. This merged and replaced the New Town Committee and the Old Town Trust. The World Heritage Trust is a company limited by guarantee with a board of directors. Unlike its two predecessor organisations, whose membership was composed of representatives of a broad spectrum of national and local organisations, heritage and community interests, the directors of the trust are appointed as individuals (Figure 7.12).

Culture of architectural conservation

Edinburgh today is marked by a strong culture of architectural conservation, a strong body of associated professional and craft skills, and a protective system that is inclusive and rigorously enforced.

Figure 7.12 Ann Street was built on the estate of the portrait painter Sir Henry Raeburn, on land to the west of the Water of Leith in the neighbourhood of Stockbridge. It forms part of the seventh and final phase of the formal developments in the New Town. Its first section was constructed between 1816 and 1825. Modestly sized by Edinburgh New Town standards, today they are amongst the most sought after houses in the World Heritage Site.

Ambition

The Bath document has been prepared to a wide remit and is predicated as a holistic management plan for the site, one that serves to link and inform other strategies, policies, and programmes for the city – including those for land use, tourism, and education – and that facilitates the insinuation of cultural heritage values into all aspects of the city's urban management and everyday life. Key words in the management plan include awareness, community, understanding, and sustainable management. A key phrase is 'supporting the local community in its cultural, social and economic vitality'. The Bath management plan is not geographically delimited: it encompasses the site as currently defined, its landscape setting, and the communities that relate to it.

The Edinburgh document is presented primarily as a framework for the conservation and management of change of the physical aspects of the historic environment, strictly within the boundary of the World Heritage Site and for the benefit of the resident and business communities that inhabit it and visitors. Key words in the management plan include conservation, development, and economic benefits. A catchphrase has been adopted: 'embracing the past, enhancing the future'.

Each of the management plans envisions that its site will become an exemplar for urban heritage management and conservation: in the case of Bath 'founded on partnerships of local, national and international communities'; in the case of Edinburgh 'using the highest standards of design and materials'.

Structure

In conformity with the indicative format provided in the ICCROM *Management Guidelines*, the two management plans are structured according to broadly the same logical sequence of description of the World Heritage Site and its significance, management issues and objectives, and management strategy and actions. Both management plans are subject to overall review every five to six years, and their action plans, annually. The latter prioritise actions according to their ongoing, short-, medium- and long-term importance, and form the basis for implementation year on year.

The Edinburgh management plan is concise and divided into sections that address landscape setting, urban form and architecture, history and heritage. It sets out a number of 'policies'; Edinburgh World Heritage does not however have a statutory role and the relationship of these to the city's local plan is ambiguous.

The Bath management plan is more wide ranging and more detailed. It sets out three over-arching objectives applicable to all aspects of the site's management: coordinated common-purpose; founded on the principles of sustainability; and sound understanding by all, including visitors. It then continues with five sections: managing change; conservation; interpretation, education and research; physical access; and visitor management.

Cultural heritage assets

The management plans adopt an inclusive approach to the cultural heritage assets of their respective sites, one that extends beyond those that fall within a strict interpretation of the relevant criteria of outstanding universal value. In the case of Bath, this inclusive approach encompasses:

- the built heritage: the buildings, streets, footways, and bridges; including the structures associated with Isambard Kingdom Brunel's Great Western Railway and John Rennie's Kennet and Avon Canal. Brunel's Paddington to Bristol railway is featured separately in the government's *Tentative List* of possible candidates for future nomination to UNESCO for inclusion on the World Heritage List;
- the parks, gardens and cemeteries;
- the archaeology: including the potential for future discoveries;

Figure 7.13 Dean Village, Edinburgh. The management plan for the World Heritage Site draws attention to the 'hidden treasure' of the valley of the Water of Leith and its mill communities.

Figure 7.15 (above) No 17 Heriot Row, Edinburgh New Town. Bath and Edinburgh share strong associations with prominent individuals in the arts, sciences and public affairs, and these are recorded throughout the historic areas of the two cities.

Figure 7.14 (left) Edinburgh has a long tradition of statues and monuments featuring famous people; also, this intimate statue of Greyfriars Bobby in the Old Town. The Edinburgh management plan promotes continuity of this tradition of public art.

- the natural environment: including the importance of the site's geology and biodiversity; and
- the intangible associations and traditions, including the culture of worship, bathing and healing associated with the hot springs; as a place of social interaction especially in the eighteenth century; and the city's associations with many prominent individuals in the arts, sciences and public affairs.

The Edinburgh management plan adopts a parallel approach, drawing attention to the 'hidden treasure' of the valley of the Water of Leith and its mill communities, and to the many statues and monuments that are such a feature in the public realm (Figures 7.13–7.15). Highlighted additionally are Edinburgh's historical and present day role as Scotland's capital city, the many intellectual, educational, professional and political associations, and the importance of the several arts and science festivals.

Principal urban management issues

Principal amongst the urban management issues affecting the two cities today are traffic and transport, development pressures in the historic centre, visitor management, and the relationship between the historic centre and the wider city.

Physical access, traffic and transport

The Bath management plan addresses the issue of physical access head-on. Whilst admitting that it is one of the most challenging and difficult to resolve, it identifies it as 'fundamental to the comprehensive management of the whole World Heritage Site', for its impact on the condition and conservation of the historic environment, people's ability to navigate and understand it, the effects of air and noise pollution on people and the environment, and the city's viability.

The environmental capacity of the city in terms of through and local traffic is addressed, as are numerous proposed improvements over time in the integration of different transport modes, prioritisation for pedestrians and cyclists, and meeting the variety of mobility needs (Figure 7.16). The evolving provisions in the city for 'park and ride' have proved especially successful in reducing city centre road access – and consequent demand for parking – by commuters and shoppers alike.

The most destructive elements of the Buchanan proposals for a new hierarchy of city centre roads were dropped in the 1970s, since which time the construction of the M4 motorway some distance to the north of the city has relieved much of the east–west through traffic. Two major regional routes still pass through the centre of the World Heritage Site, however, and a resolution for these has still to be agreed upon.

Figure 7.16 Bath: view from Milsom Street along Green Street. The Bath management plan supports the reduction of traffic in the World Heritage Site, integrated transport for the city and its surrounding communities, and further pedestrianisation.

The Edinburgh transport saga: a long, involved, and continuing story

The historic public transport infrastructure

The construction of Edinburgh's formerly extensive suburban railway system began in 1831 and reached its maximum extent in 1903. Developed by the competing North British and Caledonian Railway Companies it served all parts of the then built up area, even if in a somewhat uncoordinated manner. From the early-1940s through to the mid-1950s most of the suburban lines were closed; in 1969 the Waverley Route south to the Scottish Borders was also closed. By the early-1970s only the east–west mainline railway remained in use for passenger traffic together with a handful of suburban stations along it.

Through the nineteenth century horse-drawn buses were introduced. These were superseded first by horse-drawn trams, then by a 40-kilometre network of cable-operated trams. The first electric trams, powered from overhead wires, were introduced in 1905. They operated extensively across Edinburgh until abolished by the city council in the mid-1950s.

Pre-eminence of the motor vehicle

Largely since the mid-1950s therefore, all mechanical conveyance of goods and people within the city – whether by public or private transport – has been motorised and has competed for space on the city's roads.

The 1949 Abercrombie plan for Edinburgh was the first of an ever-lengthening list of abandoned projects for the construction of new motorway standard roads and, increasingly since the mid-1970s, a variety of schemes to revive the suburban railway network, reintroduce trams, construct an above and below ground metro, or any combination of these. Edinburgh, Scotland's capital city with a population little short of half a million, does not even have a rail connection to its international airport.

In 1986 the operation of bus services was deregulated by government, allowing competition between different companies along all routes. Although the initial chaos resulting from this measure subsided and the city council bought back its own bus company, the medium term effects were simple: people were discouraged from using the only available form of public transport and transferred to using their cars. The position has now partially recovered, the bus network has expanded, attracts more custom, and is reported to run at a profit. At the same time, the overall number of journeys by road has also greatly increased.

By the 1970s it had been recognised by many that new office and retail developments should be directed away from the city centre in order to establish a balanced planning and transportation strategy for the city as a whole. Leith to the north and Gyle to the west were particularly identified, and expansion to these sub-centres has since occurred. Nevertheless, strong encouragement has also been given to continuing major office and retail developments in the historic core, greatly increasing the number of journeys into and out of the city centre. The private car is the preferred means of transport for many people; only two narrow shopping streets, one very short, are fully pedestrianised.

Faced with the avoidable but inevitable congestion on the roads, the city council decided that the only solution was the introduction of road pricing. A referendum was called in 2005. The result? An overwhelming 'No'. The suggestion is that this was not so much a statement in favour of the motor car as defiance against well over half a century of failed plans and policies.

> ### Forwards or backwards?
>
> First, bus lanes were tentatively introduced; then, in 1997 *greenways*. Today, they accord priority to buses along principal radial routes throughout the city. Additionally, priority routes for pedestrians and cyclists have been developed – including along disused suburban railway lines that had previously been earmarked for possible reuse for public transport.
>
> Second, the first New Town especially has been subjected to a succession of traffic management schemes: all have been introduced on an experimental basis; some have only lasted a few months.
>
> Third, there are currently two schemes being promoted for new track-based public transport networks: a tram system for the city north of Princes Street, possibly with an extension westwards to the airport, and a light railway for the southern half of the city. The two schemes are being promoted by rival companies, employ different technologies, and are not compatible with each other.
>
> ### Missed opportunity
>
> Early in 1972 Arnold Hendry, Professor of Civil Engineering in the University of Edinburgh, published a proposal for an integrated rail-based public transport network to serve the entire city. This would have reused existing and disused railway tracks, with some adjustments to serve the city's expansion since 1903 and to incorporate an extension to the airport. In 1972 the city council was still employing the ubiquitous Colin Buchanan and Partners to conclude a road-based solution for the city's future transport needs.
>
> ### A twist in the tale
>
> In 1993 Jan Gehl, who has contributed so much to resolving Copenhagen's city centre planning and traffic strategy (see pages 125 to 126), was awarded the Sir Patrick Abercrombie prize for exemplary contributions to town planning by the International Union of Architects. He also holds an honorary doctorate from Heriot-Watt University in Edinburgh.

Traffic and transport are also major issues in Edinburgh and a series of experimental road schemes has been introduced across much of the World Heritage Site, with concomitant loss of visual and physical coherence, especially within and between the various planned phases of the New Town, and limited benefit to non-road users. The Edinburgh management plan does not mention environmental capacity and addresses the whole issue only in so far as it leads to 'loss of quality of townscape'.

The reality is that both city centres would benefit from robust management of vehicular movement, perhaps on the model of Copenhagen, with long-term incremental reductions in access and parking, and – especially in the case of Edinburgh – greatly increased prioritisation for pedestrians.

Development pressures

Bath has a dynamic city centre and the Old and New Towns of Edinburgh are that city's centre. The two management plans seek to address the issue of managing change positively, representing it both as a threat and an opportunity.

The Bath plan emphasises more strongly the presumption in favour of preserving that city's historic fabric and complementing it only with the highest quality in contemporary design in any new developments and infill (Figure 7.17). It also sets out the potential that high quality interventions offer for improving the condition, presentation, accessibility and use of the site, and identifies the tools that may be used to achieve this. The local plan for the city complements this by focusing pressures for the expansion of retail floorspace into less sensitive areas south and west of the line of the medieval walled city, towards the railway station and the western riverside.

The Edinburgh management plan is predicated as offering 'a positive approach in which development and conservation are seen as part of a single planned process', and cites the 1987 ICOMOS *Washington Charter* in support of this (see pages 13 to 14) (Figure 7.18). The Edinburgh local plan and related strategy documents focus development pressures into the very heart of the World Heritage Site, especially for expansion in the retail sector, for which a figure of 50,000 square metres of additional floorspace has been

Figure 7.17 The city of Bath has long had a reputation for the poor quality of its modern architecture, much of it characterized as neoclassical pastiche. The Bath Spa Building (built 2000–06; Nicholas Grimshaw, architect) is vaunted in the city as a statement of confidence in contemporary architecture and the extension into the twenty-first century of Bath's tradition of technical and architectural innovation.

Figure 7.18 Until the 1990s, Edinburgh also had a reputation for conservatism with regard to modern architecture. The new Scottish Parliament (built 1999–2004; Enric Miralles, architect) is vaunted in the management plan for the Edinburgh World Heritage Site as an exemplary twenty-first-century addition to the city's architectural heritage.

quoted. The Edinburgh World Heritage management plan is complemented therefore by proposals for major redevelopments in parts of the first New Town – certain of which would involve the demolition of listed buildings – and the construction of high-capacity underground car parks in both the Old and New Towns. All of this will, of course, increase pressure on the transport infrastructure.

Edinburgh World Heritage was established in the same year as one of its main partners, the Edinburgh City Centre Management Company. Edinburgh World Heritage describes the City Centre Management Company as a 'sister organisation'. It is a private sector led partnership whose remit is to promote the economic success of the city centre: as the retail focus of the city, for business and property development, and through visitor spending. The Management Company seeks to expand the catchment area for the city centre's retail sector well beyond Edinburgh's administrative boundary, and opposition has been expressed to retail developments in communities up to 50 kilometres distant – which are perceived as a threat to the city's expansionist ambitions.

Writing in 1990, the Prince of Wales described Edinburgh as 'the most beautiful and civilised city in Britain'. Speaking at a conference organised by Edinburgh World Heritage in 2006, the Prince is quoted as follows: 'There is a real and present risk that, in the drive to make Edinburgh a world city in the commercial sense, we make it more and more like just any other city in the world, and in so doing, diminish its status as a beacon of excellence in architecture and urbanism, and indeed enlightenment'.

Visitor management and relationships to the wider city

The Edinburgh management plan does not consider any relationships with the wider city, its heritage assets and its communities, except for issues affecting the townscape and landscape setting of the World Heritage Site itself. Whereas, for example, the Bath management plan seeks to diffuse the present focus of visitors away from the main city centre attractions, thereby easing the physical pressures and distributing the economic benefits around the city-wide community, the Edinburgh management plan seeks to reinforce the visitor focus on its city centre, including that the city's various festivals should continue to be concentrated in it.

Focus on the Edinburgh World Heritage Site has exacerbated longstanding perceptions of socio-economic divisions in the city. These have been expressed by the author J K Rowling in her introduction to *One City*, published in 2005. Writing of her time in the city in the 'rags' phase of her life, before Harry Potter, she writes of the 'barriers, invisible and inflexible as bullet-proof glass, that separate those in the affluent, able-bodied mainstream of our society from those who, for whatever reason, live on its fringes'. She goes on to refer to the abyss between the 'violence, crime and addiction [that] were part of everyday life' in her part of the city, 'barely ten minutes away by bus [from] a different world, a world of cashmere and cream teas and . . . imposing facades'.

These divisions persist: one has simply to use the park and ride service from the south-east side of the city to witness them – or venture to the social housing in the north-west of the city. Despite its affluent image, one in five of Edinburgh's children is reported to grow up in a household below the income support level.

The Edinburgh management plan neither acknowledges nor addresses these realities of the World Heritage City that is Edinburgh today. As if – albeit unwittingly – to emphasise this, the World Heritage Site management plan gives encouragement to the demolition and redevelopment of buildings of lesser architectural quality in the site simply on design grounds. Does this echo Viollet-le-Duc or the *Burra Charter*? And how does it relate to the *Brundtland Report*? The World Heritage Site may be a jewel, but where is the crown?

By way of contrast, an important initiative in the Bath management plan is the promotion of the benefits of World Heritage Site status to the various local communities across the city, including in deprived areas: benefits economically, socially, and as a valuable tool for learning, culture and leisure.

Conservation responsibilities

The two management plans set out the various instruments and guidance that underscore the protective systems that apply to their respective sites. The Edinburgh management plan relies on these. The Bath management plan recognises that statutory protection through the planning system is insufficient on its own. First, because the United Kingdom consent process for the historic environment is first and foremost a paper exercise and is not monitored after the issue of approvals. Second, because there is now no local authority supervised grant scheme in the city that would encompass the inspection of actual works.

The Bath plan proceeds to highlight the responsibilities for maintenance and conservation that attach to the many individual owners across the site, and promotes engaging with the community to create a sense of common ownership of the conservation ethic: a culture of excellence in which good practice becomes the norm.

Bath and Edinburgh are widely recognised for their exemplary professional and technical skills in architectural conservation. The two management plans underscore the need to secure continuity of these skills and the availability of appropriate building materials from sustainable sources. They also promote strategies for preventive maintenance.

Resources

In the United Kingdom, as we have seen, World Heritage Site status does not carry with it either additional statutory protection or financial aid.

In Bath, a forty-year historic buildings repair grants programme – funded jointly by central and local government – came to an end in the financial year 1995/96. It has not been replaced. Indeed, even the single post of World Heritage Officer has, to date, only been secured on an annual basis. The management plan informs and coordinates the discharge by Bath and North-East Somerset Council of the full range of its local authority functions, but there are no year-on-year budgetary allocations that are specific to World Heritage Site issues. The Bath management plan strongly represents that World Heritage Site status carries additional responsibilities.

By contrast, Edinburgh World Heritage inherited continuity of the substantial funding that was enjoyed by its two predecessor organisations, employs eight full-time staff, and currently manages an annual budget of 1.25 million pounds for its conservation funding programme.

Monitoring

The formats for cyclical reporting that are set out in the UNESCO *Operational Guidelines* anticipate that the state of conservation of World Heritage Sites will be monitored on the basis of key indicators.

The monitoring indicators that have been adopted by Edinburgh World Heritage are mostly statistical and include: number and status of listed buildings, buildings at risk, proportion of site covered by conservation areas, number of planning applications, population, retail indicator, visitor numbers and spend, and financial assistance distributed. These are to be collated and compared on an annual basis.

There is, as yet, no formal monitoring programme in place for the Bath World Heritage Site. The management plan proposes a broadly parallel set of key indicators, but includes the impact of planning applications as well as their quantity and considerations such as: number of historic buildings with conservation or management plans; number and result of historic building condition surveys; and the negative impact of permitted development rights.

Statistical approaches are rather simplistic. The Bath management plan acknowledges the need to devise a set of monitoring indicators that relate more closely to the expectations of the UNESCO *Operational Guidelines*, especially in respect of the importance of safeguarding authenticity and integrity – for which the first step is to carry out a baseline audit of the whole site. Such an audit has yet to be carried out.

Chapter 7: digest

The various and several UNESCO, ICCROM, and ICOMOS documents coupled with the parallel global agenda of sustainable development raise many expectations with regard to the management of complex cultural heritage sites and their relationship to socio-economic activity in the modern world. The devising of holistic, heritage-led management plans for historic cities is a new field. To embrace, additionally, the sustainability agenda is charting new territory not simply for the United Kingdom but internationally.

In the context of cities that are on the World Heritage List, of which there are now over 200 worldwide, there is a lack of precision in the guidance. The international documentation does, however, set down clear markers as to the range of issues and the appropriate responses. These are not comprehensive, but they represent a sound starting point.

The Bath and Edinburgh management plans have been drafted to different remits. The plan for the Bath World Heritage Site has the ambition to be a coordinating, heritage- and sustainability-driven management plan for the city. As such, it poses more questions and explores more answers. Although there is much in that plan that has yet to be tested, and the dedicated resources are very limited, it is a broad-ranging and visionary document that has relevance to historic cities worldwide. It has the significant advantage of addressing a whole city – both its historic environment and its communities – and being authored by the principal, governmental organisation with responsibility for implementing it.

By contrast, the plan for the Edinburgh World Heritage Site is essentially a stand-alone framework for conservation and development in the historic centre. It conforms to the limited expectations of established United Kingdom practice and does not chart the way forward for holistic awareness or skills in management. It marries with the development-led ambitions of Edinburgh World Heritage's partners, fails to flag up the risks that are attached to overheating in the commercial property market, and has no apparent relationship to any coherent and sustainable one-city vision.

Many of the limitations of the Edinburgh plan suggest that it is unhelpful to attempt to draft a management plan for a historic city centre in isolation from its urban context. Also, that Edinburgh World Heritage needs to adopt a more robust approach to safeguarding continuity of the intangible cultural heritage values of the site as well as the tangible ones.

It is ironical that when Bath developed as a classically planned city in the eighteenth century it achieved this in part by superimposing on the medieval city, whereas today it is focusing expansion in the retail sector in less sensitive parts away from the historic core. In Edinburgh on the other hand, its particularity in the eighteenth century – one that is specifically recognised in the inscription of the Old and New Towns as a World Heritage Site – was the harmonious juxtaposition of the two highly distinctive parts, but new retail development in the city today is being overlaid on this historic core, as are the car parks to serve it.

In short, whereas the Bath management plan is underscored by the principles of sustainability, the Edinburgh plan appears to confront them.

Chapter 8
Managing Historic Cities: the Bottom-Up Approach

Central and Eastern Europe

A wealth of historic cities

The Central and Eastern region of Europe boasts a wealth of historic cities that complements that of the Western region and greatly extends the range of its influences: climatic, cultural, religious, and historical.

Characteristic of the historic cities that escaped extensive damage in the 1939–45 War – the vast majority – is the remarkable degree of survival through the socialist period, to the extent that many of the challenges facing historic city centres in Central and Eastern Europe post-1990 have been similar to those that were confronted in Western Europe following the First and then the Second World Wars. Of these, most relevant, is a building stock that was generally in sound structural condition and in use notwithstanding decades of neglect and a lack of modern amenities. Characteristic also is the predominance of mixed-uses in historic core areas: housing intermingled with small shops, markets, workshops, and commerce, all with the human culture and contact that accompanies them.

Housing

Housing is the predominant building use in historic cities throughout the region. It is the use for which the majority of the buildings were constructed and it remains the use for which they and the urban infrastructure are all best suited. Housing is the starting point for their revitalisation.

In Budapest, for example, programmes of restoration and upgrading to the many hundreds of tenement blocks have been a political priority for a number of years, and grants, loans, and professional expertise are made available by the municipal authorities to groups of individual proprietors to facilitate such works (Figure 8.1). In Hungary as a whole there is a particularly well-organised tenants' association that has proved to be an effective lobby for the interests of occupiers – whether tenants or owners – of tenement flats.

Figure 8.1 Budapest, Hungary. Programmes of support for the restoration and upgrading of tenement housing operate in the city centre districts on both sides of the river Danube.

In Poland, housing rehabilitation policy, practice, and funding mechanisms that were developed in France from the 1960s onwards have been introduced and proved highly successful in certain cities. Funding mechanisms include underwriting and subsidy agreements between municipalities and local banks, directed at offering very low or interest-free loans to tenants to acquire their properties and to support their contributions to conservation and rehabilitation schemes. Achieving the effective management of the housing stock and empowering a stakeholder society are principal aims of this policy.

Sibiu, Romania

The city in its region

Sibiu was founded by Saxon settlers in the twelfth century. It is also known by the historical name of Hermannstadt and it retains strong cultural and other links with present-day Germany. Historically the cultural, political, and religious centre of Transylvania, its historic core is one of the largest in Romania: the extent of its outer fortifications covers 86 hectares; within this area, its resident population is 14,000. It has an urban landscape that is dominated by vernacular architecture rather than individual monuments. Unlike the Romanian capital, Bucharest, its historic centre survived unscathed from the presidency of Nicolae Ceausescu (from 1974 to 1989).

Sibiu is noted for its uninterrupted traditions of ethnic and functional diversity and it remains the focus of cultural and commercial life for its region.

Figure 8.2 Sibiel, Romania. Sibiu is at the centre of a region whose ethnic diversity stretches back several centuries; clusters of Hungarian, Romanian and Saxon villages are to be found in all geographical directions. Sibiel in Sibiu County, a Romanian village, is typical of rural Transylvania.

The city and its surrounding area are not, however, by Western European standards, prosperous. As with much of Romania, Sibiu is located at the heart of a region of limited economic activity, where the infrastructure is poor, where subsistence farming is the norm, and where the horse and cart are the means of everyday conveyance for the majority of the population (Figure 8.2).

Romania signed the World Heritage Convention in 1990. The historic centre of Sibiu was placed on the Tentative List for Romania in 2004.

Romanian–German Cooperation project

In 1998 an international conference was held in Sibiu under the auspices of the United Nations Educational, Scientific and Cultural Organisation (UNESCO) and the Council of Europe. This identified threats to the cultural heritage of the city, such as neglect and the absence of coherent policies, and established the need to define and promote a positive vision and a coordinated programme for conservation and sustainable development (Figure 8.3). This conference focused local, national, and international attention on the city and led to the initiation of an urban rehabilitation programme led and part-financed by the German Agency for Technical Cooperation, GTZ (Deutsche Gesellschaft für Technische Zusammenarbeit). From the outset, this programme has been based on a place-specific, bottom-up approach to the rehabilitation of the historic core.

At an early stage of this Romanian–German Cooperation project a comprehensive study was undertaken of the housing conditions, the social

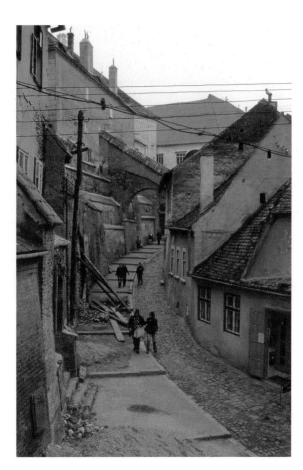

Figure 8.3 Sibiu, Romania, was founded by Saxon settlers in the twelfth century. Following decades of neglect, the historic inner fortifications have now been the focus of a major programme of consolidation and reconstruction (in 2002).

composition, and the views of the inhabitants in the historic centre. This established that sixty per cent of the housing was owner occupied, only nine per cent was fully renovated, and over fifty per cent lacked basic amenities or was in a poor state of repair – with a significant proportion of residents sharing toilets, bathrooms and kitchens. Space standards were low by Western European standards, a high proportion of residents were in the low or very low income brackets, and there was a bias towards the elderly and the retired compared to the overall city population – all of which reflects experience elsewhere.

Significantly, eighty-five per cent of the residents stated a clear preference that their dwelling be improved, rather than be obliged to relocate to more modern accommodation elsewhere in the city. Another important finding was of a strong self-help ethic amongst owners and tenants, who had both the experience and the willingness to participate in carrying out repair and renovation works themselves and to share their skills with neighbours and relatives.

The conclusions of this and other wide-ranging studies were consolidated into the *Charter for the Rehabilitation of the Historic Center of Sibiu/*

Figure 8.4 Piața Mare, Sibiu's main square, has been cleared of traffic and repaved as part of the traffic management and public space improvements in the historic centre (in 2006).

Hermannstadt, published as a consultative draft in March 2000 and finalised in October 2000. This Charter argues that conserving both the tangible heritage and the living character of historic Sibiu are fundamental to preserving its identity for future generations and to securing a sustainable future for it.

The Charter's objectives and priorities cover a full range of topics and issues: from service infrastructure and housing; through cultural tourism and retail; to townscape, open spaces, and traffic (Figure 8.4). The key mechanisms for achieving this are a strategy and action plan prepared by GTZ on behalf of the City Hall, the latest being for the period 2005–09. This defines and establishes roles, tasks, and the funding to be provided from the local community; local, regional, and national funds; statutory undertakers; and international contributions.

The over-arching aim of the GTZ project is to build local capacity for urban rehabilitation: the ethic, the people, the institutions, the tools, and the finance. Attracting a significant level of outside private sector investment is not a primary objective of the strategy nor does it depend on it. Activating and making best use of available resources within the local community is, thereby securing socio-economic continuity. Gentrification – such as has occurred in many Western European cities – is not, at least at present, an issue. Gentrification requires an economy that supports gentry who wish to live in a historic centre: Sibiu does not have such an economy or class of people.

Through its local office in the heart of the historic centre, GTZ contributes to and manages a grant-aid budget for local residents to carry out external and internal conservation works and upgrade their houses or flats. Additionally, it manages and funds the provision of initial free professional counselling and publishes an extensive range of technical advice leaflets and booklets that promote best conservation practice. Priority is attached to a holistic understanding of the historical evolution and the environmental performance of buildings, and awareness of these is widely publicised in the community.

Importantly, both from the architectural conservationist's viewpoint as well as financial necessity, GTZ advocates and supports local residents in the implementation of a careful, gradual, and economical approach to rehabilitation, one that respects significance and historical layers. Grant-aid calculations, based in principle on a ceiling of fifty per cent, take into account contributions in kind by owners and tenants who undertake works themselves, essential in an area of predominantly low-income households who would not otherwise be able to afford to have works carried out. Importantly also in a historic core where the vernacular predominates, no distinctions are drawn between whether individual buildings are listed as monuments or not (Figures 8.5–8.7).

Other key actions include support in the training of local professional and craft skills, and the setting-up of small-scale workshops in the city centre, complete with associated residential accommodation. Sustaining diversity in the historic centre through the vertical mix of functions in buildings is an important strategic objective.

Figure 8.5 View from the upper to the lower town in the historic centre of Sibiu. Vernacular architecture predominates, and residents are supported in the rehabilitation of their homes irrespective of whether their properties are listed as monuments or not (in 2002).

Managing Historic Cities: the Bottom-Up Approach 167

Sibiu: *Avram Iancu street*.

Sibiu: *courtyard of housing*.

Figures 8.6 and 8.7 The housing in the upper part of the historic centre of Sibiu is typically arranged as apartments around open courtyards, the apartments on the upper levels being accessed by open galleries. Avram Iancu street is one of many streets in which the apartment blocks have benefited from the GTZ supported programme of conservation and upgrading (in 2003).

Figure 8.8 Sibiu, Romania. The restoration of the exterior of this restaurant on the corner of Piața Mică received an award in 2002.

A shortage of minimum intervention adaptive skills has been identified amongst professionals – especially architects and engineers – and an ongoing programme of seminars and workshops seeks to address this. It is a problem that relates to the core training in the construction professions. It is not of course unique to Romania.

Very much a bottom-up project, the starting points have been the existing material and human resources: the urban plan and the buildings, the people who use and occupy them, and the socio-economic community that they form (Figure 8.8).

Community involvement in the historic centre

In 2002 at a conference held at Ironbridge Gorge, England, Dave Askins, a lecturer at the Telford College of Arts and Technology, spoke of programmes aimed at engaging with young people. Through his experience with the Telford Schools World Heritage Project he stressed the importance of involving young people as active participants in their heritage. Treated as passive consumers, their interest in heritage is not awakened. Treated as participants, they are amongst its most valuable ambassadors.

Early in 2006 a campaign was launched in the historic centre of Sibiu aimed at informing and persuading residents not to use PVC as a substitute for the traditional joinery of doors, windows and shutters. It is a campaign that included engaging with school children through a competition in which they were invited to submit written work, artwork, and performance. The level of involvement and the standard of the entries were such that the Mayor decided to host a celebratory barbecue in the main square and himself took part in

Community involvement elsewhere in Romania

Community involvement is also a feature of other tangible-intangible cultural heritage projects and programmes in Romania. Aimed at all sections of the national community, whether by age or ethnicity in what is a pluralistic society, young people form a special focus.

Banffy Castle

At Banffy Castle, Bonțida, near Cluj-Napoca – a project that involves the restoration of the castle and landscaped grounds of a former aristocratic estate that survived the Second World War only as a ruin – the involvement of school children in replanting the park, and in

Figures 8.9 and 8.10
Banffy Castle, Bonțida, Romania. The restoration of the castle and park is being managed by the Transylvania Trust together with the Institute of Historic Building Conservation. A multi-faceted project, it is having a major impact on the revival of traditional craft skills and engages with all age groups and sectors of the local community.

consequence becoming custodians of it, has played an important role in integrating the project into the community (Figures 8.9 and 8.10).

The overall project involves all generations. Significantly, it has provided many training and employment opportunities in the restoration of the various structures and in the establishment of new businesses, including guesthouse accommodation. Additionally, it has engaged with the ethnic diversity of the area. The involvement of the local gypsy community in the festivities on 'Bonțida Cultural Days', with horse riding, singing, and dancing, has proved a particular success.

Schools project

A nationwide schools project, initiated by a non-governmental organisation in Brașov and supported at an early stage by the British Council in Romania, is taking tangible and intangible heritage issues into the heartland of school curricula. It is rooted in the principle of participation through interactive teaching and community involvement, not in treating pupils and students as passive consumers strictly within the classroom environment.

the cooking. Mayor Klaus Johannis – an inspirational figure in the local community – has spearheaded the anti-PVC campaign, including signing the promotional leaflets.

Asmara, Eritrea, Horn of Africa

Context

In 2004, I was asked to coordinate and conclude the work of a number of national and international consultants, non-governmental and governmental organisations, in the preparation of over-arching urban planning and building conservation guidelines for Asmara's historic core.

The tangible cultural heritage of the city is young by European standards; its intangible cultural heritage is at least as old. The geographical, historical and human context is specific – as it is in any established settlement and community that has evolved over time. The starting point for the whole exercise has been the place and its people.

Eritrea

Eritrea owes its geographical footprint as well as its name to Italy, the last of the European powers to join the 'Scramble for Africa' in the nineteenth century. By purchase and conquest the Italians established Eritrea as a colony in the 1890s. They chose an ancient name for it, Eritrea being derived from the Greek word for 'red'. The country's Red Sea coastline is over 1,000 km long.

Modern Eritrea officially declared its independence from Ethiopia in 1993. The estimated population of between three-and-a-half and four million inhabits a land area that is slightly larger than that of England. The country's economy is very largely one of subsistence agriculture, some of it worked by

Figure 8.11 Eritrea. Only twenty-five per cent of the population lives in urban areas.

nomadic tribes, and only twenty-five per cent of the population lives in urban areas (Figure 8.11).

Eritrea has been home to people of diverse living patterns, traditions and religions for thousands of years, and today's population comprises nine ethnic groups and two main religions: Christianity and Islam.

National identity and sustainable development

Concern for the preservation of Eritrea's cultural heritage in its many forms, tangible and intangible, was a major preoccupation before the country gained independence and this has accelerated since. Cultural heritage is seen both as an essential component of affirming and promoting national identity and a cornerstone of sustainable development. As such, conservation is being harnessed to economic development initiatives, including cultural tourism.

The Cultural Assets Rehabilitation Project (CARP) was formally established in 1997, with governmental and popular support; the World Bank is its major partner. CARP's mandate is to identify, preserve, and promote awareness of the diversity of Eritrea's cultural heritage, and it is engaged in a number of parallel programmes across the country. CARP's projects include archaeological sites, record management and museum development, and the recording of oral history and traditions. The built environment is one of CARP's key programmes and Asmara is one of its projects.

Asmara

The Italians moved their colonial capital from the Red Sea port of Massawa to Asmara in 1899. Situated on the Eritrean Highlands at an altitude of 2,400 metres, it is Africa's highest capital city. The area has been settled since the eighth century, when four distinct villages were established. In the fourteenth century they merged to become 'Arbate Asmara', which approximately stands for 'the four united'. The construction of European building types – as opposed to the modest, stone-walled and either flat- or conical-roofed vernacular dwellings – dates from 1889, the year in which the Italians first occupied the area.

From the 1890s onwards the principal areas of the city were planned – as colonial settlements frequently are – to an expanding series of orthogonal grids, adjusted to suit historic caravan routes and natural features. The later and more developed city plans from the Italian period show spacious European quarters in the southern part, a denser mixed quarter centred on the area of the main markets and mosque in the northern part, and an industrial quarter in the north-east corner. Organically developed indigenous quarters were kept outside this planned city.

Asmara was envisaged primarily for an emigrant working-class population from the homeland, which provided the skilled labour force to consolidate the sub-Saharan colony. The new European quarters were laid out to the model of a garden city, with wide tree-lined boulevards and residential streets – planted with a limited range of species including palms and jacarandas – and low-density housing set behind hedges dense with bougainvillea. Asmara provided an urban ideal far from the cramped and unhealthy conditions of nineteenth-century cities in Europe. In mainland Italy, the 1930s' town of Sabaudia, built on the reclaimed Pontine Marshes 80 kilometres south of Rome, is analogous.

Until the early-1930s the architecture of the new city displayed an eclectic range of historical styles, to which architectural historians have attached a multitude of labels: Classical, Romanesque, Medieval, Moorish, Islamic, Renaissance, Lombardian, Venetian, Neo-Gothic, Neo-Baroque, Neoclassical, Alpine, Colonial, Italian vernacular, indigenous vernacular, and early-twentieth-century Novecento.

The greatest period of construction in Asmara occurred in the six short years between 1935 and 1941, coinciding with the span of Italy's conquest of Ethiopia. The building boom transformed the predominant historicism of the city's early architecture into 'Africa's Secret Modernist City', an urban environment that is unique in Africa and has few parallels anywhere in the world. The legacy of neglected, but largely untouched, Modernist, Rationalist, and Futurist buildings is recognised as one of the great architectural survivors of the twentieth century (Figure 8.12).

Eritrea signed the *World Heritage Convention* in 2001. The 'Historic Perimeter of Asmara and its Modernist Architecture', as the extent of the Italian colonial capital is known, was placed on the Tentative List for Eritrea in 2005.

As the capital city of a young nation, there are many pressures to encourage development. At the same time, there is a strong sense of the unique identity of Asmara and of the historical role played by Eritrean labour in its construction and evolution.

In 1999, pending the preparation and agreement of over-arching urban planning guidelines, the municipal authorities imposed an embargo on

Figure 8.12 Asmara offered a blank canvas that permitted uninhibited architectural expression. One of the Futurist icons of the city is the Fiat Tagliero service station (built 1938; Giuseppi Pettazzi, architect). The design was inspired by that of an aeroplane: the cantilevered reinforced concrete wings protect the petrol pumps from the heat of the African sun.

virtually all development and change within the 4.0 square kilometre historic core. The area of the historic core was defined by CARP, who researched and identified some 400 historically important buildings within it. CARP has also collated archival material on a total of over 800 buildings across a wider area of the city.

It is not, however, just the built heritage whose importance is recognised. Since 1941 Eritreans have adopted the city as their own, adapting themselves to it and vice versa.

The legacy of Italian customs is manifest, especially in what were the European quarters: the café culture, complete with cappuccino, macchiato, and expresso coffee; Italian cuisine with pastas, pizzas, and pastry shops; the two-hour lunch break; the *passeggiata*, the leisurely evening stroll up and down Harnet Avenue, where Asmarinos go to look, be seen, and meet their friends (Figure 8.13); and the cherished and ubiquitous Fiat Cinquecento. Aspects of the city's life are pure Mediterranean transposed at altitude on to African soil.

But this is only one aspect of Asmara's rich and varied living culture, of the fusion of twentieth-century Europe, Italian Modernism, and a disparate indigenous culture.

The area of the main markets, with their countless stalls selling every imaginable item of food, clothing, craft goods, and with an all-pervading scent of spices, express a *mélange* of Eastern and African influences (Figure 8.14).

The industrial area of the Italians' planned city, Medeber, is yet another aspect of Asmara's rich hybridity. Here, hundreds of self-employed stallholders

Figure 8.13 Asmara's principal thoroughfare, Harnet Avenue, is the setting every evening for the leisurely promenade known as the *passeggiata* – one of the many legacies from the Italian era.

apply incredible ingenuity and resourcefulness to recycling every imaginable material: rubber tyres into rope and sandals; oil cans into cookers. It is the artisan workshop of the city, without which the basic needs of a large proportion of its citizens would not be met (Figure 8.15).

Figure 8.14 Asmara. The elongated market square, with its arcaded stalls along each side and shared covered spaces between. The main market area is very extensive and covers several dozen small-scale city blocks.

Figure 8.15 Asmara: Medeber, the small-scale industrial area to the north-east of the city centre, strategically placed between the railway station and the market square, is enclosed within the old caravanserai. Medeber accommodates hundreds of artisan stalls. Most of the recycling work that is carried out in this area is done by hand without the use of modern machinery.

And finally, there are the indigenous quarters that have evolved organically, beyond the main markets and Medeber to the north.

Fusing the needs of architectural conservation with those of continuity of this varied living culture, not for reasons of sentimentality but as an essential part of the social and economic functioning of the city, is the united objective of the various governmental, non-governmental, international and national partners who are working together in Asmara. CARP has been leading this process, seeking to harness the complex individual elements into a single programme of sustainable development.

Strategic plan for the city

Under the colonial administration, Asmara's Italian population expanded from 3,500 in 1934 to 55,000 in 1940. In 1941, following the defeat of the Italians in the Horn of Africa, the British counted 60,000 Italians and 100,000 Eritreans in the city.

Today, the population of the city is estimated to be over 450,000. Taking into account natural population growth, projected migration from rural into urban areas, the return of expatriates from abroad, the pre-eminence of Asmara in the urban hierarchy of the country, and other factors, the population of Asmara is expected to double in size – approaching one million – by the year 2015 and to experience continuing growth thereafter.

As the capital city, strategically located at the hub of international and national transport networks, Asmara will be the focus for increasingly heavy demands for floor space for large-scale commercial activities – offices, shops, and hotels – and the expanding administrations of the national government and municipal authority.

Conventionally, in many countries of the world, these pressures have been focused in the historic hearts of cities, with serious destructive consequences to their built environments, social balance, craft and artisan industries, and evolutionary development.

Successive local and international consultants, the Municipality of Asmara, and CARP are agreed that the best interests of the capital's historic centre will be served if it is allowed to evolve in tune with its tangible and intangible cultural heritage rather than be subjected to destructive pressures for major redevelopment. Asmara is seen to require an approach that supports complementary developments elsewhere in the wider city to enable it to meet the new and expanding needs without compromising the historic core.

A strategic plan has been prepared that seeks to diffuse pressures for large-scale commercial development to a limited number of sites well away from and out of sight of the city centre. One such site is at Sembel, close to the airport, comprising the Expo exhibition centre, the Intercontinental Hotel, an established village, and a newly completed European-style development of apartment housing complete with integral shops and community facilities. The principle is analogous to that of an *urban village*.

Skyline and urban landscape

Historic Asmara is located at the centre of a small depression in the high plateau, surrounded by a ring of hills.

Traditionally, its skyline has been dominated by the symbols of the ethnic and religious diversity of its population. Most notable within or immediately adjacent to the historic centre are the campanile of the Catholic cathedral; the minaret and domes of the Grand Mosque; the square towers of the Orthodox cathedral; and the spire of the Protestant church (Figure 8.16). These and others are the landmarks that orientate the local population as well as visitors, and which give visible expression to the cultural heritage and diversity of the city.

This skyline was transgressed on a few occasions between the early-1970s and early-1990s. The 1999 embargo on new development prevented further high-rise construction in the historic centre, despite pressure from some private investors to demolish and rebuild on existing and vacant plots.

Under the Italian colonial administration, development in the city was strictly controlled. A maximum height of 16 metres – stipulated as four

Figure 8.16 The skyline of historic Asmara is dominated by the towers and spires of its churches and the minarets of its mosques.

storeys – was permitted for buildings in the commercial heart. Generally across the city a two-storey maximum height was imposed, with limited discretion for three storeys.

A block-by-block study of the historic area has been carried out and guidelines prepared that generally restate the original planners' intentions. The existing urban landscape of the historic core is less densely built up than the Italians envisaged, and the effect of these guidelines is to encourage the more efficient use of buildings and land and to support an extensive number and range of development opportunities without destroying the city's overall image and scale. An extension of height controls across a wide surrounding zone is also proposed.

Land and building uses

Although historic Asmara was divided into European and mixed quarters, it was never zoned by use except in the designation of the industrial quarter. Throughout the central area there is a complex mixture of uses horizontally by plot and vertically by floor. Even in the areas of one- and two-storey residential villas, small factories, workshops, and shops exist cheek-by-jowl with the housing, all easily accessible by foot (Figure 8.17). Asmara is very much a lived-in and liveable city, with many different layers of social and economic interaction, most of it unplanned and informal. The streets are alive with human activity.

Asmara has been described as representing an ideal that urban planners all over the developed world are trying to reintroduce into cities.

Figure 8.17 A joinery workshop in one of Asmara's predominantly residential areas.

The urban planning guidelines for the historic centre do not seek to sanitise it or to rationalise the complex socio-economic relationships by imposing any system of zoning by use, except that certain activities have been classified as incompatible: by type – heavy industry; and by scale – large offices, retail stores, and hotels. There are well over a thousand small businesses in the historic area that provide employment for several thousand of its residents. Retaining them in their present locations is seen as essential both to their survival and to the diversity and vitality of the heart of the capital. Implementing this aspect of the guidelines will require refined planning tools that can support mixed housing and small-scale offices, retail, guest houses, and workshops.

The industrial area of Medeber has been the subject of an in-depth socio-economic and physical planning study for the Ministry of Trade and Industry. This proposes a sensitive ten-year phased programme that is aimed at rationalising the use of the land and buildings; providing opportunities for technical training and career advancement; and improving the general environment. These latter include partial pedestrianisation and the provision of landscaped open spaces. This study sets the standard for a people-orientated, methodological approach that is relevant to other sensitive parts of the city.

Traffic and transport

Studies have been carried out by academics and consultants from the Netherlands and Sweden aimed at resolving issues in the historic centre, including noise and air pollution, through traffic, escalating car ownership, and lack of parking spaces and parking controls. Detailed proposals aimed at curtailing through traffic and the use of private vehicles, and prioritising public transport, cyclists and pedestrians have been coordinated with the urban planning guidelines.

Housing

Housing in the historic centre ranges from exclusive villas – several now in use as embassies and ambassadorial residences – through modest single-family houses and apartment blocks, to cramped courtyard housing in the low-rise market area (Figures 8.18 and 8.19).

The population of the 4.0 square kilometre historic centre is estimated to be 35,000; research has confirmed that the Italian planners anticipated a population of 50,000 within the same overall area. This indicates a present-day shortfall of 15,000, and maintaining and wherever possible increasing the present number of inhabitants is seen as essential to the city's sustainable development. The opportunities for development that have been identified in the urban landscape studies have been coordinated to enable the Italian planners' intentions to be realised.

Space standards and basic facilities in the area vary enormously. The opportunities for development also facilitate the improvement of space standards and living conditions at the affordable end of the market.

Figure 8.18 The entrance to one of Asmara's several larger villas. Dating from the 1910s, it is distinctively European.

Figure 8.19 Asmara. A typical access path serving the low-rise high-density courtyard housing in the market area.

Public open space

Colonial Asmara was conceived as a green city, but it has lost much of its public open space, street planting, and other greenery in recent decades. There is also a severe shortage of defined public open space in the form of parks and play areas.

It is a characteristic of Asmara that streets in residential areas are part of the public open space. They are the places where adults meet and children play, both regularly and informally: children under the constant and watchful guard of parents, neighbours, and passers-by. Asmara is an Eritrean city, not a leafy middle-class suburb in Western Europe where such activities would be frowned upon and discouraged. The principal conflict – whether actual or potential – is with motorised vehicles.

Generally, it is proposed that opportunities will be taken throughout the city centre to enhance tree planting in the public domain: in the main boulevards and in streets in the residential and markets areas. Additionally, extensive use of traffic-calming measures and home zones together with additional hedge planting is proposed, thereby securing large parts of the public domain as safe and colourful community space.

Tentative use of public art has already been employed in the historic area, in the use of blank walls for depictive painting and roundabouts for sculptures. The plan is that the use of public art will be extended throughout the historic core, including interactive types such as sculptures for children to climb and benches for adults to sit on. A public art policy will serve to extend the city's cultural heritage and provide a shop window for the nation's creative artists.

Historic buildings

The over-arching urban planning framework is intended to take the heat out of pressures for large-scale destructive redevelopments, thereby securing the socio-economic and environmental conditions for an evolutionary approach that supports architectural conservation. The complementary building conservation guidelines adopt the *Burra Charter* as their starting point: they emphasise the importance of understanding and safeguarding significance, minimum intervention, and the control of the rate of change.

A particular concern relates to original interior fittings and furnishings – mirrors, light fittings, bar counters, stools, tables and chairs – which many of the publicly used buildings such as cafés, theatres, and hotels retain *in situ* (Figure 8.20).

The design of new buildings

The Italian administration's 1938 Building Regulations set out a number of precepts concerning building heights, distances of setback from street frontages, together with a number of details, but did not attempt to restrict architectural style. In a city that displays remarkable stylistic variety within

Figure 8.20 The Selam Hotel (built 1937; Rinaldo Borgnino, architect) is vaunted as the finest example of 1930s' Rationalist architecture in Eritrea. It remains in its original design state, inside and out, including its fittings and furnishings.

a very limited time span, restrictive design guidance appears inappropriate. Rather, it is proposed that design continuity should be encouraged and monitored. The vehicle for this is the Committee for the Historic Perimeter of Asmara, which advises the Municipality on urban and conservation issues.

Chapter 8: digest

Much can be learned from the experience and potential of historic cities where a top-down approach based on preconceived norms is neither relevant in theory nor capable of implementation in practice.

The urban conservation and sustainable development project in Sibiu offers insight into how the material resource value of cultural heritage harnessed to bottom-up community involvement can drive urban conservation in situations where forced change and exploitation of property assets for their development value is neither desirable nor practicable. Progress with the project has been impressive, as has the revival of confidence in the historic centre on the part of residents and businesses alike.

In Asmara I encountered a determination on the part of the various governmental and non-governmental partners to interpret a national and local cultural vision into a coordinated environmental, social and economic programme for the city, eager to maximise the diverse tangible and intangible cultural heritage values of the place and its people in all their complexity. The over-arching urban planning and building conservation frameworks are intended to incorporate key elements of sound practice in sustainable development – all as perceived and interpreted by a range of organisations and interests working to common purpose.

Much has still to be done in the city. Eritrea is a new nation, lacking many of the legal and administrative structures with which others are familiar and which they take for granted. But other countries, with all of these ostensible advantages, have often squandered the inheritance of their historic environments and the human cultures that accompany them. Asmara is seeking to take heed of these lessons of misspent wealth and is working hard to avoid making the same mistakes. It is not a city that has the wealth to misspend on self-destruction.

Both Sibiu and Asmara are working to an agenda of enhancement through evolutionary processes that engage with the principle of minimum intervention: environmentally, socially, and economically. Their historic environments are the catalyst and inspiration for this.

Chapter 9
The Coincidence between Conservation and Sustainability

Think global, act local

Sustainability

Environmental concerns have been expressed increasingly and from a number of complementary directions since the 1950s. They have been reinforced by a series of highly publicised individual disasters that have had direct consequences on human lives, and by cumulative actions by man that have resulted in progressive damage to the Earth's protective ozone layer and change in the climate. A key aspect of this awareness raising has been communication of the understanding that all life forms on our planet form part of the same, complex, mutually supportive ecosystem.

The 1987 *Brundtland Report* established that current patterns of resource consumption and environmental degradation cannot continue, and that economic development must adapt to the planet's ecological limits. The 1992 Rio Declaration set out the need to conserve, protect and restore the health and integrity of our ecosystem.

The *Brundtland Report* popularised the term *sustainable development*: the proactive component of the continuous over-arching process of *sustainability*.

Sustainability embraces a number of concerns. These include loss of habitat and biodiversity; escalating consumption of non-renewable material and energy resources; and pollution, emissions of wastes, and their relationship to the health of the planet and its biosphere.

Sustainability recognises that the historical equilibrium between humans and the natural environment has been seriously disturbed at all levels, from the global to the local, that the recovery of balance is essential, and that the twin starting points are the international community as a whole and each individual in his or her community.

Sustainability exhorts a constructive evolutionary approach, one that prioritises human development on the basis of equity between peoples and generations – notably improvements in health, education, and quality of life – over economic growth focused on raising the already affluent lifestyles of a few.

The '3 Es' of sustainability: economy, environment, and equity (also expressed as: energy, environment and ecology)

Seven environmental concepts

The sustainability debate has accumulated a number of catchphrases and some of the best known are brought together in the headings within this chapter. The debate has simultaneously gathered a series of environmental concepts, of which the following are the most relevant in connecting sustainability with conservation in the context of historic cities.

Environmental area

An urban area within which considerations of the environment predominate over the use of motorised vehicles and through traffic is excluded. It may also be applied to considerations relating to people.

Environmental capacity

The volume and character of traffic in a street or area that is compatible with the maintenance of good environmental conditions. As a generic concept, *environmental capacity* has far broader relevance for historic cities than the single issue of traffic in streets. It may also be applied to determine the capacity for pedestrians; also, the capacity of the buildings themselves (Figure 9.1).

Figure 9.1 San Gimignano, Italy, in 1966. This picturesque hilltop city covers an area of 45 hectares. Its narrow streets now receive over three million visitors a year; it has been estimated that on peak days there are two tourists per square metre. The congestion and pollution resulting from traffic and the erosion caused by pedestrians have become major causes of concern. San Gimignano is considered to have exceeded its *environmental capacity*.

Environmental capital

In the general sense: the stock of natural resources and environmental assets. It is applied to reflect the embodied materials and energy that have been invested in and are represented by existing buildings and urban infrastructure.

Environmental impact

The impact of any action on the use or misuse of *environmental capital*. Buildings constructed and maintained using renewable, indigenous materials have low *environmental impact*; likewise, if they are energy efficient in use.

Environmental performance

The manner in which the structure of a building performs in relation to its external and internal environmental conditions – including thermally and by relative humidity. Traditionally constructed buildings are said to 'breathe', and their *environmental performance* is an expression of this. They adapt to their climate rather than exclude it. Built using the materials available in their locality, they maximise the use of free ventilation, cooling, heating and daylight, all to match their climatic conditions. Performance standards in use can generally be raised by low impact maintenance and modifications at a fraction of the environmental and financial costs of major alterations or reconstruction.

Environmental responsibility

This is self-explanatory.

Environmental stakeholder

The status of each and every individual, community and society that assumes *environmental responsibility*.

Sustainability emphasises the essential relationship between biodiversity and cultural diversity, between the specific characteristics of each and every natural environment and the human living patterns that inhabit or otherwise relate to them, and that loss of equilibrium in either risks unsettling both.

Sustainability comprehends that cultural diversity is an essential component of cultural identity, sense of community belonging, social inclusion, and participation.

Conservation

Cultural identity expresses itself in many ways and, as we have seen, is categorised in the context of cultural heritage as tangible and intangible.

Architectural conservation has evolved into a broad discipline that recognises geocultural diversity and local distinctiveness, specifically as they are expressed through the physical identity of places, buildings, and architectural

Montepulciano: *Window*.   Urbino: *Window*.

Figures 9.2 and 9.3   The cities of Montepulciano and Urbino in Italy are only 146 kilometres apart. Geocultural diversity and identity are reflected clearly in the architecture of their buildings, the materials of construction and the craft skills employed.

details – at all levels from the monumental to the vernacular (Figures 9.2 and 9.3).

The safeguarding of authenticity and integrity – however they may be defined in each and every location – are prerequisites for the continuity of both the tangible and the intangible components of cultural diversity and all that they imply in terms of societal identity and cohesion (Figure 9.4).

A key message that arises from an analysis of the background and theory of architectural and urban conservation is minimum intervention – also expressed as control of the rate of change: to the fabric of buildings, the urban grain of historic cities, and the socio-economic structures that inhabit them.

This conservationist message resonates with the analogies drawn by Sir Patrick Geddes and Gustavo Giovannoni to the natural sciences, and to the consonant emphasis in the definition of the *sustainable city* that realisation of the concept depends on the city being viewed and managed as an ecosystem, thus taking us back full circle to the awareness raising resulting from the environmental concerns.

This conservationist message also resonates with the United Nations Educational, Scientific and Cultural Organisation (UNESCO) agendas that

Figure 9.4 Braşov, Romania. The geocultural identity of a historic city is expressed in a multitude of ways. The urban landscape affords the overview: here at Braşov, of a medieval walled city with its towers and Gothic church – the largest of its period in Romania – set protectively into the surrounding mountain range.

increasingly emphasise the importance of intangible heritage and an anthropological vision of culture as a dynamic and evolving process in which the narrow concept of heritage as relics and records from and about the past is superseded by one of socio-cultural continuity and enhancement (Figures 9.5 and 9.6).

In this, historic cities are perceived not so much as static objects to be admired for their history and architecture, but as living spaces to be occupied and appropriated by local communities as an essential part of the process of safeguarding those communities' identity and sense of belonging. The preservation of material objects maintains its place but is augmented by reproduction, transmission and adaptation by present and future generations.

Continuity of the demand for traditional building craft skills and their extension into multiple new areas is just one of numerous positive outcomes of this anthropological approach, contributing to positioning heritage as an integral part of socio-economic and cultural life today not as peripheral to it. It represents a significant expansion of the relevance and potential of architectural and urban conservation and of the means to engage in them successfully.

Reduce, reuse and recycle

The '3 Rs' of non-renewable resource and waste management form an essential part of the coincidence between conservation and sustainability.

Athens, Greece: *the Erechtheum (built 421–407 BC)*. Cheltenham, England: *Montpellier Walk (built 1842–43)*.

Figures 9.5 and 9.6 As we have seen, authenticity may be a defining concept but it is no longer viewed as a restrictive one either in time or space. Neither is culture itself – but as 'a process and a negotiation of connections'. Architecture gives tangible expression to these connections, as may be seen by relating the caryatids of the Erechtheum on the Acropolis at Athens with those dividing the shops in Montpellier Walk in Cheltenham more than two millennia later.

It is reported that the United Kingdom construction industry accounts each year for the use of 6 tonnes of building materials per head of population and thirty-five per cent of the total of all wastes. It also accounts for the manufacture of 3.5 billion bricks and the destruction of 2.5 billion. 'It doesn't', as one commentator has written, 'take a genius to work out the absurdity of such an equation.'

Such images of large-scale consumption of resources and production of wastes have also to be related to reports that, worldwide, the human population is currently burning one million years' worth of fossil fuels every year. Whereas it may be considered that cities in the poorest countries of the world are the most sustainable, in that everything that can be is recycled and food is produced in inhabitants' backyards, supermarket chains in the United Kingdom have now taken note and, for example, started introducing compostable packaging manufactured from renewable vegetable sources.

The sustainability argument reinforces the view that the historic environment should no longer be perceived in limited cultural terms only for its archaeological, architectural and historic interest. The rationale for a conservationist approach is greatly enhanced when the cultural significance of the historic environment is allied to its environmental capital, at all scales up to and including the historic city. This does not imply a halt to development or reduction in the activity of the construction industry. Rather, it implies a reorientation of it focused on new development that is additive and complementary, and significantly increased emphasis on maintaining, reusing, adapting, and enhancing the existing built stock and infrastructure – all within an overall framework that embraces the principles of the sustainable city and coordinated urban management.

Disparities such as the demolition of soundly built dwellings in the North of England, the non-use of vacant upper floor premises in historic city centres, and the use of greenfield land in the South-East of the country for massive programmes of new house building, do not accord with the sustainable management of either cultural or material resources.

The conservation-sustainability approach is not new. Indeed, historically, pre-industrialisation, it was the norm in all civilisations. Building materials were recycled and buildings reused; an evolutionary, additive process was taken for granted; the material resource value to individuals and communities was the primary motivation; and top-down academic interpretations of cultural significance had not been formulated and played no part. Post-industrialisation, strategic approaches such as that promoted by Gustavo Giovannoni, pursued in his native Italy, and applied at the metropolitan scale in Paris, reflect and continue this pre-industrialisation norm.

In Britain today, materials are recycled but at a limited scale. The recycling sector tends to focus on higher value architectural features and not on basic construction materials.

The record in reusing buildings has greatly improved since the 1970s, numerous publications have highlighted the opportunities that historic buildings offer, and a raft of television programmes and associated publications has encouraged the do-it-yourself market and designer-makeovers in the modernisation of older housing.

The foundations of the knowledge and enthusiasm are all there. The gaps however are significant, and they include the absence of coherent national strategies, coordination at the urban scale, and interdisciplinary understanding and skills. They also manifest an absence of a regard towards historic buildings individually, and cities holistically, that prioritises minimum intervention over unnecessary levels of disruption to fabric, environmental performance and communities alike, and focuses on complementing them with additive development.

At the same time, one of the legacies of consumer-orientated societies is to regard buildings as disposable objects that have a limited lifespan. 'It would seem reasonable', as one publication from the late-1960s stated, 'that dwellings should be replaced after a maximum life of sixty years.' Built-in obsolescence has characterised much post-1945 construction in Britain. In the late-1980s, immediately prior to the 'crash' in the commercial property market in central London, the design life for some newly constructed speculative office blocks was as low as fifteen years, and the prospect loomed that they might never be occupied.

Recycling building materials within the community

South-East Scotland

The recycling of building materials is a time-honoured tradition that dates back to the age when mankind first started making structures and has been customary in all civilisations. Coincident with the decline of its Empire from the third century onwards, Rome became one huge recycling yard. Materials from all of the major monuments – the Coliseum to take just one example – were removed and reused in other buildings.

The Scottish Borders contains the ruins of four medieval abbey complexes – Dryburgh, Jedburgh, Kelso and Melrose – all of which were destroyed in warfare in the mid-sixteenth century. Stonework taken from them can still be identified in domestic and other buildings across their local communities (Figure 9.7).

My own house, restored 1980–83, contains recycled stone, bricks, slates, wood, joinery items, ironmongery and other fittings that were either available on site or recovered from a number of buildings that had been or were in the process of being altered or taken down (see Figure 5.6 on page 90).

Figure 9.7 Dryburgh Abbey, Scotland. From the date of its founding in 1150 to its demise in 1544, the abbey complex was destroyed by fire on three occasions and ravaged by war on four. Stonework salvaged from the abbey can be seen in a variety of structures in the locality.

Reusing buildings within the local community

Melrose Station, Scotland

The former station house at Melrose was completed in 1849. It was designed by John Miller, one of Scotland's early railway engineers. Melrose was the gateway to an emergent tourist industry that was inspired by the references to abbeys, castles and the natural landscape

Melrose Station: *new restaurant.*

Melrose Station: *derelict interior.*

Figures 9.8, 9.9 and 9.10
Melrose Station, Scotland. One of only two top grade historic monuments in the town – the other being the ruined medieval abbey – the rapid dilapidation of the former station house following the closure of the line in 1969 became a *cause célèbre* in the local community. The restoration focused on providing the building with new functions within that community. (Acquired 1985; restored 1985–86; managed together with the adjacent former cycle shop and garage as a mixed-use commercial development 1986–2003; Dennis Rodwell, architect/developer.)

Melrose Station: *restored exterior.*

in Sir Walter Scott's novels. Listed in the top category of historic monuments, Melrose was described as 'the handsomest provincial station in Scotland' when it was opened.

The station formed part of the Waverley Route from Edinburgh south through the Scottish Borders to Carlisle in the North-West of England. The 160-kilometre-long route was closed in 1969 and the building rapidly fell into ruin. By the mid-1980s it was the only town station still standing on the line and had become home to scores of pigeons (Figure 9.8).

Constructed on the side of a hill, the building served as the stationmaster and signalman's houses at the lower level, and accommodated the ticket office and waiting rooms at the upper, platform level.

The rescue and reuse of the station, together with adjacent buildings on the same site, focused on providing them with a function in the local community, and offering flexible commercial spaces that have served a variety of purposes – including craft workshops, offices, medical consulting rooms, crèche, furniture showroom, retail space, and restaurant – all within a range of sizes to suit the locality and the businesses that set up or moved into them (Figures 9.9 and 9.10).

From the mid-1970s through to the mid-1990s major, multi-million pound programmes of tenement housing refurbishment in Edinburgh were based on prolonging their lifespans by thirty years, a period that has now begun to elapse. Constructional techniques were applied to extensive tracts of the inner city that only had this limited objective in view; they led to considerable damage to external masonry, specifically through the use of short-term stone repair and cleaning methods.

The life of structurally sound buildings is as long as we choose to keep and maintain them properly. It has no other limit.

Reusing buildings within the local and international communities

Bolshoi Theatre, Moscow, Russia

The Bolshoi Theatre is important not so much for the architecture of its showcase building as for its symbolic status as the focal point of a 3,000-strong community of musicians, dancers, singers, designers, technicians and administrators that together make up one of the world's leading companies in the performing arts – notably in ballet and opera.

The Bolshoi has been described as a perfect example of the linkage between tangible and intangible heritage, inherited culture and continuous creativity. It is a community of multi-talented individuals who work as a team, and there is said to be an almost spiritual linkage between the theatre buildings and their community.

The main theatre dates from 1856 and incorporates elements of its two predecessors, both of which burned down. The interior, its safety measures, its machinery, and its electrical equipment were substantially unaltered since the 1930s (Figures 9.11 and 9.12).

In 1993 a UNESCO-Bolshoi project was established with the view to securing a comprehensive refurbishment programme that would balance best building conservation practice with maximum functionality to the standards expected of an international artistic venue today.

Bolshoi Theatre: *exterior.*

Bolshoi Theatre: *interior.*

Figures 9.11 and 9.12 Bolshoi Theatre, Moscow, Russia is the subject of a major three-year programme of upgrading following several years of impasse between conservationist and modernists.

In 2003 a conference of technical specialists and theatre directors from across Europe met in Moscow to resolve the impasse that had arisen between uncompromising conservationists and equally uncompromising modernists.

In 2005 the theatre was closed for a major three-year long programme of restoration, partial reconstruction, and upgrading. Emphasis has been placed on securing continuity of the intangible traditions of the theatre company, including the informal relations and mutual respect between all members – from the prima ballerina to the novice seamstress.

Reusing an industrial complex: minimising intervention and maximising potential within the community

Darley Abbey Mills, Derby, England

The emerging cotton industry that inspired the development of the factory village of Darley Abbey was established by Thomas Evans in 1782. The use of the site for textile manufacture ceased in 1970, since when the buildings have housed a number of engineering and light industrial enterprises. Significant parts of the complex have now been underused or vacant for several decades.

The city of Derby is best known as a nineteenth-century railway town, for its Crown Derby bone china and Rolls Royce aerospace factories, and for its football team. The significance of the Darley Abbey industrial village was only recognised in the local community in the year 2000 when – along with Sir Richard Arkwright's Cromford, and Jedediah Strutt's Belper and Milford – it was nominated as part of the Derwent Valley Mills World Heritage Site.

The former cotton manufacturing complex forms part of the Darley Abbey conservation area. The majority of its buildings are now listed, several in the top category. The entire complex – from the largest textile mills with their clear internal spaces, through the cart sheds, workshops, and mill manager's house, to the stables and gatehouse kiosk – are complete and predominantly in their original state (Figures 9.13 and 9.14).

In 1991 the Civic Trust proposed a regeneration programme for the site. It was not sympathetic to the architectural and historic importance of the complex; it did not proceed.

Darley Abbey Mills awaits a coordinated revitalisation programme that respects its architectural and historic significance and provides a viable long-term future for the entirety of

Figure 9.13 Darley Abbey Mills, Derby, England. The mills depended initially on the water-power of the river Derwent to drive the machinery for textile manufacture. This photograph shows, left to right: the North Mill (in the distance; built c.1825); the canteen (centre foreground; built 1820s); Long Mill (the oldest and tallest; built 1782–83, then remodelled 1789–90 following a fire); and the boiler house chimney (built 1896).

Figure 9.14 Darley Abbey Mills: the raw cotton preparation range and workshop (built in two phases, 1790s to 1811).

the complex. Such a programme is most likely to be successful if it focuses upon uses that relate to the local community and adopts a minimum intervention approach to the technical aspects of conservation. This will safeguard the significance of the site, be the least costly, and maximise its potential for reuse.

Stay close to source

This catchphrase of sustainability is underscored by the embracing concept of *proximity*, whether it be of place of work to place of residence, of education to leisure, or, importantly for architectural conservation, of traditional building materials and craft skills to the localities in which they were employed historically and for which they are best suited today. Reduction in the need for travel and transport for everyday purposes, and the unnecessary use of non-renewable energy sources in the process, is a key beneficial consequence.

Top-down meeting bottom-up

One of the key lessons from the twentieth century is the contrast between the assumption that man could explain and order everything, and the realisation that human society and the natural world are far more varied and complex than many people had thought and hoped.

Top-down theories in urban planning in particular failed to provide the results that had been forecast. The ordering of cities according to separation of functions and programmes of slum clearance and community dispersal generated a series of land use, transport movement and social problems that were not manifest previously and which we have inherited.

As Jane Jacobs wrote in 1961:

> Cities happen to be problems in organized complexity, like the life sciences.... The variables are many, but they are not helter-skelter; they are inter-related into an organic whole.... This is a point of view which has little currency yet among planners themselves, among architectural city designers, or among the businessmen and legislators who learn their planning lessons... from what is established and long accepted by planning "experts". Nor is this a point of view that has much appreciable currency in schools of planning (perhaps there least of all).

'Good planning', wrote Donald Insall a few years later, 'is only good management.'

Top-down solutions in urban planning seek to impose received ideas that often originate in the abstract on real life situations for which they are frequently ill-suited. At the same time, in today's societies it is inclusive top-down mechanisms that offer the means by which bottom-up solutions may be realised.

Bottom-up solutions start from analysis and understanding of the identity of a historic city both in terms of the continuous evolution of its tangible heritage and of its human culture. They make specific demands on planning and building professionals to work with what exists and not to seek to impose incompatible received ideas and technical solutions. They enable informed choices to be made about what is significant in a historic city and for local distinctiveness to be safeguarded through common ownership by its community.

In urban planning terms, bottom-up solutions allow the buildings, the plot sizes, the street patterns and open spaces, together with the traditional patterns of use, movement and human interaction to determine the least interventionist approach to the environment, the society and the economy of a city. They afford a solid foundation upon which to secure *conservation* and *sustainability* jointly, as one inseparable process. This is the ambition of the practical model of Sibiu and the theoretical model of Asmara.

Bottom-up approaches in historic cities do not seek to sanitise them of certain classes of people, categories of use, or types of existing building that from an aesthetician's point of view do not subscribe to ideal urban stage sets. They accept, if not welcome, the organised chaos of historic cities, warts, colour and untidiness included. Most importantly, they recognise that cities neither exist nor function without citizens (Figures 9.15 and 9.16).

The conjunction between top-down and bottom-up in the over-arching process of sustainability is a matter of mutual understanding and skill sharing – of empowering citizens and communities to act as *environmental stakeholders*. This is the key message that filters from the 1992 United Nations Rio *Earth Summit* through to Local Agendas 21.

Figure 9.15 Zamość, Poland. Market squares were designed to serve as places of commercial interchange and human interaction. There is a tendency in some historic cities to relocate markets away from their squares on the premiss that they make the architecture and townscape look untidy.

Figure 9.16 Paris, France. The *bouquinistes* along the banks of the Seine are an essential part of the human culture of the historic centre of Paris.

In many respects the top-down, bottom-up conjunction aims to recover the pre-industrial understanding and respect between mankind and his natural habitat.

Figures 9.17 Banská Štiavnica, Slovakia, is a medieval mining centre that developed into a Renaissance city and became the most important centre for the extraction of precious metals during the Austro-Hungarian Empire. The remarkable degree of survival in historic cities across Central and Eastern Europe is explained by their value as a usable resource rather than for their architectural and historic interest.

Urban conservation: expanding threats to historic cities

Misspent wealth in development

This book begins by quoting the architect-planner Graeme Shankland:

> Today in most western countries it is the mis-spent wealth in development which is the biggest agent of the destruction of historic cities, not physical decay.

The remarkable degree of survival, generally, of the historic environment across Central and Eastern Europe throughout the post-Second World War socialist period may be accounted for quite simply: first, the economic and ideological pressures for redevelopment and replanning were absent; and second, it was too valuable in terms of its usefulness for it to be destroyed in the name of progress. Historic cities across the region may not have been well looked after, but at least they survived (Figure 9.17).

Shankland wrote these words in 1968, but there is a strong sense in which his message remains valid in Western countries and that the syndrome has become contagious.

Figure 9.18 Prague, Czech Republic, was placed on the World Monuments Fund Watch List as a consequence of threats posed by overheating in the development market.

Prague, Czech Republic

Following the political and economic changes, the development market in the Czech capital became overheated. Much of the authenticity and integrity of a historic centre that had survived the 1939–45 War and forty-five years of socialism was threatened to the point that the city was placed on the World Monuments Fund Watch List of 100 most endangered sites for a period in the 1990s (Figure 9.18).

Sighişoara, Romania

Sighişoara, a medieval citadel deep in rural Transylvania and reputed as the residence of Vlad Dracul in the years 1431–35, was threatened with the development of a *Draculand* theme park on the high plateau barely two kilometres distant (Figure 9.19). The plateau was partly cleared. Surplus army ammunition stores on the side of the hill were used to dynamite 600-year-old oak trees. Huge billboards announced: 'Here we are going to construct Dracula Park'.

The project was abandoned in 2003. The World Heritage Site itself lacks a meaningful management plan.

St Petersburg, Russia

Across Central and Eastern Europe during the socialist period there were few examples of proactive urban conservation as opposed to the high-grade

Figure 9.19 Sighișoara, Romania. The citadel was inscribed on the World Heritage List in 1999. UNESCO led an international campaign to block the development of *Draculand* on the plateau above the city and the project was abandoned in 2003.

restoration and sometimes reconstruction of selected individual historic monuments (Figures 9.20 and 9.21).

St Petersburg was founded by Peter the Great in 1703 and was the capital of the Russian Empire until the 1917 Revolution. The city centre and surrounding historic sites were inscribed on the World Heritage List in 1990. The city as a whole covers 200 square kilometres; the population is five million.

The conservation zone of the city centre conforms to the extent of the city up to 1910, covers 55 square kilometres and includes 3,500 historic monuments (Figure 9.22). The urban landscape of the city is controlled by a high-buildings policy that embraces a protected area that extends substantially beyond the conservation zone.

During the Soviet era the polluting industries in the city region – steel, chemical and petrochemical – were perceived as the major threat to the historic environment.

The years between 1991 and 2004 saw significant changes in the demographic profile of the conservation zone. Its population dropped from 900,000 to 500,000; offices supplanted residential accommodation; and the concentration

Catherine Palace at Tsarskoie Selo: *exterior view of the Catherine Palace at Tsarskoie Selo* (commenced 1717; predominantly 1752–56; Francesco Bartolomeo Rastrelli, architect; several of the interiors were remodelled in the 1780s by Charles Cameron, architect).

Catherine Palace at Tsarskoie Selo: *interior of the Great Hall.*

Figures 9.20 and 9.21 Tsarskoie Selo, near St Petersburg, Russia. The three best known imperial country palaces in the surroundings of St Petersburg – Petrodvorets to the west, and Pavlovsk and Tsarskoie Selo to the south – were substantially destroyed towards the end of the Second World War and subsequently reconstructed with meticulous attention to detail. They epitomise the scientific skills in the restoration of major monuments that were characteristic across much of Central and Eastern Europe through the socialist period.

Figure 9.22 St Petersburg, Russia. The triumphal double arch in the General Staff Headquarters building, linking Palace Square to Nevsky Prospekt (1819–29; Carlo Rossi, architect).

Figure 9.23 St Petersburg, Russia. Housing rehabilitation is one of many challenges in the city. The hundreds of imperial and aristocratic palaces are not considered adaptable to modern residential use: only to offices or institutional purposes. A number of apartment blocks in the city were built in the second half of the nineteenth century when the custom was to rent rooms, not flats. A dozen or more families may share a single kitchen and washroom.

of movement by public and private transport in the city centre greatly increased. Built on a swamp, there are major engineering difficulties in extending the metro system and increasing the frequency of its stations.

Pressures for redevelopment in the conservation zone are increasing, such that it has been proposed to reduce its size to a fifth of its declared extent, thereby encouraging projects of redevelopment across the major part of it. The area that would remain would encompass the principal historic sites, the main canals, and lengths of the banks of the river Neva. This is perceived as offering 'a satisfactory compromise between maintaining the city's historical face and stimulating property development'.

In principle it is intended to maintain height controls, but the policy changes will involve a site-by-site, project-by-project, decision-making process. This, in the absence of an over-arching land use, transportation strategy, and an increasingly flexible approach to the urban landscape. Even maintaining the city centre population at the level of half a million may prove a challenge too far (Figure 9.23).

Shanghai, China

It was reported in the early-1990s that fourteen out of China's largest fifteen cities retained sufficiently strong urban agriculture sectors as to render them virtually self-sufficient in food. This is a position that has altered significantly in the intervening years of major redevelopment and expansion in cities such as Shanghai, a city that was surrounded by some of the most fertile land in the country.

The city has expanded from a population of 5 million in 1949, through 7.2 million in 1957 and 13 million in 1990, to 20 million today. By 2020, Shanghai's population is expected to reach 30 million. Questions are being asked about the world's most populous country's continued ability to feed itself.

One of the most polluted and congested cities in the world, bicycles have now been banned from some areas of the city centre as they interfere with the movement of private cars, and several of the most recently built tower blocks have a design life of fifteen years.

Exporting the threats

It would seem that Western countries have exported problems rather than solutions.

Chapter 9: digest

Sustainability has spawned a number of catchphrases, and several of these are used in this chapter: think global, act local; the '3 Es'; reduce, recycle and reuse; stay close to source; and top-down meeting bottom-up.

These catchphrases help to trigger consciousness of key principles and the actions that flow from them. In turn, they give rise to concepts that have direct application in practice, such as environmental capital, environmental performance, and environmental stakeholder.

Together, they support understanding of the impact that the actions of individuals and communities have on environmental concerns at the global level, and how those actions may be adjusted in order to relate more closely to common objectives. Above all – within the context of this book – they emphasise the actual and potential areas of coincidence between conservation and sustainability at all levels, from the scale of a single recyclable brick to the destiny of major cities.

They also serve to highlight areas of dissonance, and the environmental risks and costs that are involved in situations where the post-1945 lessons of misspent wealth in development in historic cities in Western countries are ignored.

Chapter 10
The Challenge and the Opportunity

The purpose of this book

As set out in the Introduction, the purpose of this book is to contribute to the forging of linkages between architectural and urban *conservation* and the broader environmental agenda of *sustainability*.

By juxtaposing their separate origins, seeking to make coherent sense of the complex and sometimes contradictory theoretical backgrounds, and outlining their parallel evolution and development, the earlier chapters have sought to raise many of the issues, pose a number of questions, make some of the connections, and suggest favourable directions.

The various and diverse examples of practice, from the scale of the individual building through to the metropolitan city, are all intended to insinuate real-life challenges and opportunities into what would otherwise be an essentially philosophical treatise. These examples exhibit widely different methodologies – even within the same cities over relatively short spans of time. All of which serves to confirm that inclusive rather than exclusive approaches are needed: in the definition of the challenges; in the appraisal of the options; and in the evolving focus of debates and resolutions. Change, after all, is the one constant in cities.

The challenge

Sustainable world

Cities around the world occupy a fiftieth of its land surface, house fifty per cent of its population, and account for seventy-five per cent of its annual consumption of natural resources and discharge of wastes. A significant proportion of those material and energy resources are non-renewable, reusable materials are being wasted and not recycled, and toxic wastes are polluting the oceans and the atmosphere.

The city is one of the greatest challenges of the twenty-first century. It is the starting point for a sustainable world.

Sustainable cities

To meet this challenge, principles of ecological sustainability need to be insinuated into all aspects of urban planning, from the global down to the local scale. At the theoretical forefront is the concept that cities should be regarded and managed as ecosystems: mini-ecosystems within their individual localities; and interrelated ecosystems globally. A key part of this is the recovery of natural cycles of resource use and waste management.

Pre-industrial cities offer models of sustainable urban development, functioning as they did in a balanced ecological relationship within their sub-regions. The challenge is to recover the over-riding principle of balance in an industrialised world in the age of globalisation.

Historic cities

Sustainability has three components – environmental, social, and economic – of which the environmental takes precedence as it underscores the survival of all life forms on our planet.

Historic cities start with two essential qualities: first, the environmental capital that is represented by their buildings and urban infrastructure; and second, the socio-cultural values that they signify and the role that these perform in defining sense of place, community belonging and social cohesion. These socio-cultural values are represented by a continuous time line: past, present and future. They are expressed in the architecture, the urban grain, and the socio-economic organisation of cities.

The physical and societal attributes of historic cities are inseparable. They embrace environmental issues, tangible and intangible cultural heritage, and equity both within and between generations. Recognising and acting upon the full range of values inherent in historic cities is a core component of the challenge.

Architectural and urban conservation

Writing in 1975, European Architectural Heritage Year, the perception of several leading conservation architects was simple: that the starting point in a historic city must be its historic quality and visual character not – as one writer put it – 'secondary social, economic or even ecological arguments'. Also, adopting a perfectionist urban design approach, that: 'The first principle of conservation is to keep the good parts of cities and rebuild the bad parts'. Such attitudes are reflected, for example, in the 2005 management plan for the Old and New Towns of Edinburgh World Heritage Site.

Writing on the eve of the new millennium in 1999, Jukka Jokilehto, formerly with the International Centre for Conservation in Rome (ICCROM), now consultant to the International Council on Monuments and Sites (ICOMOS) and the United Nations Educational, Scientific and Cultural Organisation (UNESCO), asks 'if the conservation movement, as it evolved from the eighteenth century, cannot be considered as concluded, and whether modern conservation should not be redefined in reference to the environmental sustainability of social and economic development within the overall cultural and ecological situation on earth'. That redefinition has yet to be advanced.

Some current issues

This section focuses on some key issues and connections. It draws together several of the threads but does not aim to be exhaustive.

The concept of heritage and its role today

The construct of *heritage* as something that relates to tangible objects from the past which we may add to today by constructing the monuments of the future – characterised frequently as *iconic* buildings – needs to be substituted by an anthropological vision: a dynamic approach that is focused on processes that safeguard geocultural identity and secure its continuity. These processes embrace all expressions of interaction between the physical environment and human activity.

In relation to architectural conservation, these include training and continuity of employment opportunities in traditional craft skills in locations where demand for their services is – whether actually or potentially – concentrated. These employment opportunities can only be secured through effective policies and coordinated urban management, both of which are currently lacking.

In this, the encompassing term *historic environment* is a start, that of *cultural landscape* more helpful, and wider recognition of the inseparable relationships between tangible and intangible cultural heritage and the continuous past-present-future time line imperative.

The fact that over half the annual turnover in the United Kingdom construction industry relates to repair and maintenance work points the way to consolidating the role of conservation as a mainstream rather than a marginal activity.

Functional, material and cultural resource

The essential difference between architecture and art forms such as painting and sculpture is that buildings exist to perform a function, the range and nature of which vary enormously according to time and place, and human desires and needs both temporal and spiritual. Architecture is about creating three-dimensional spaces and enclosures within which internal and external activities may be performed – be they ceremonial, religious, educational, commercial, domestic, or simply passing the time of day. The issues of architectural style and detail, degree of ostentation or humility, are variables that contribute to defining geocultural identity. The underlying principle of functionality is invariable. The history of architecture is the history of building construction and human use.

Establishing continuity of function is the *sine qua non* of successful architectural and urban conservation, and the principle of minimum intervention to fabric and community alike favours uses for buildings individually and historic areas collectively that relate as closely as possible to those for which those buildings and areas were constructed (Figure 10.1). Minimum intervention is a principle that is shared by *conservation* and *sustainability*.

Historic buildings and areas represent a non-renewable capital resource – of materials, energy, and financial investment – as well as a cultural one. As the editor of the *Architectural Review* wrote as long ago as 1970: 'It is the mark of an immature culture – a demonstration of a childish attitude to valuable and historic buildings – to assume that if new accommodation is required . . . it can only be provided by demolishing . . . and rebuilding on the same site'.

At the scale of the historic city, continuing resource value is nowhere more apparent than across Central and Eastern Europe, where absence of redevelopment policies kept them very largely intact throughout the socialist period.

The strategic approach

The strategic approach that is demonstrated in cities such as Chartres and Paris is essentially a bottom-up approach that starts with what exists and

Figure 10.1 Facadism is technically complex, financially expensive, and constitutes a form of architectural taxidermy that treats historic cities as theatrical stage sets. It symbolises a failure to establish continuity of function and is the antithesis of a sustainable approach to historic cities. (Glasgow, Scotland.)

seeks to make it work in the modern world. It is the principle promoted by Gustavo Giovannoni of recognising multiple values, avoiding destructive superimposition, and welcoming harmonious coexistence.

At the metropolitan scale of Paris it changed a monocentric city into a polycentric one. This is the model that is favoured in the debate on sustainable cities. It offers the opportunity to establish or consolidate neighbourhoods of proximity, balance land values and transport movements, and enables freedom of architectural expression in locations that avoid conflict. The juxtaposition of La Grande Arche at La Défense and the Place des Vosges in the Marais quarter on the front cover of this book expresses this freedom of expression and counterbalancing avoidance of conflict.

It is illogical and confrontational to concentrate the most volatile commercial uses and development pressures of a modern city in its oldest and most environmentally sensitive areas. Yet, in Britain, this is precisely what we continue to do, thereby monotonising them and providing models for destructive redevelopment that seriously threaten cities such as St Petersburg, one of the best-preserved major cities of Central and Eastern Europe. Such confrontation is the antithesis of sustainable development.

Urban landscape

The integrity of a historic city's urban landscape defines its over-arching geocultural identity to citizens and visitors alike. In recent years UNESCO has been at the forefront of debates over the intrusion of high-rise construction into urban landscapes and the failure that they signify of acceptable relationships between new development and the historic environment.

UNESCO has successfully opposed high-rise development schemes for Cologne (Germany), Vienna (Austria) and Esfahan (Iran), and engaged in ongoing debate over the skyline of the City of London.

France benefits from a national protective measure that encompasses urban landscapes. No such measure exists in the United Kingdom.

Contemporary architecture and the role of architects

Just as *heritage* is a construct so is *contemporary* in the sense that is used in relation to architecture today. Etymologically the word has a double meaning: first, occurring at the present time; and second, conforming to modern ideas in style or fashion. The first goes without saying and has no preconditions attached to it. The second is less evident, especially in a pluralist society. Is there an arbiter? Is *contemporary* an inclusive concept or a recipe for confrontation?

The sequence of conservation charters outlined in Chapter 1 illustrates a variety of attempts to position conservation in relation to the Modern Movement in architecture – especially at the time of the latter's aggressive infancy, from the early-1930s. The debate in relation to historic cities is

The essence of good urban design

'The agreement to differ within a recognised tolerance of behaviour'

Geocultural identity and sense of place are rudely interrupted when modern buildings are intolerant of their neighbours and confront them abrasively (Figures 10.2 and 10.3).

Figure 10.2 Moscow, Russia. A bland high-rise block towers over a quarter of the city that is comprised predominantly of low-rise villas.

Figure 10.3 Helsinki, Finland. The white Italian marble of the rectilinear elevations of a company headquarters (built 1960–62; Alvar Aalto, architect) snubs the revivalist architecture of its neighbouring late-nineteenth-century church.

Urban landscape and geocultural identity

The post-Second World War reconstruction of Nuremberg, Germany

Nuremberg was all but destroyed by aerial bombardment. Its post-War rebuilding sought to recover key aspects of its socio-cultural identity: through its urban landscape, the restoration of major monuments, and the reconstruction of the historic core as a mixed use community (Figures 10.4–10.6).

Figure 10.4 (left) Nuremberg, Germany. The historic urban landscape was reinstated in the post-War rebuilding of the city.

Figure 10.5 (below left) Major monuments were restored and new buildings constructed alongside.

Figure 10.6 (below) The historic street pattern and mix of uses were recovered, and the new buildings were designed to offer contextual continuity rather than historical replication.

Figure 10.7 Kutna Horá, Czech Republic: the former Jesuit College. A high level of creativity and skill is needed in order to work with as opposed to against historic building complexes where there is no longer a demand for their original use. There is a shortfall of creative skills in minimum intervention within the construction professions.

unresolved, and there is no 'agreement to differ within a recognised tolerance of behaviour' (see page 21).

Architecture exists to perform a function of internal and external enclosure. It is also a civic art. Part of the problem today is that the architectural profession is divided into three distinct classes: modernists, revivalists and conservationists. Notwithstanding that the word *architect* simply means *master builder*, the first of the three categories consider themselves the elite, the second are dismissed as practitioners of pastiche, and the third are sometimes not appreciated as architects at all. It is an unnecessary debate, fuelled by insecurity and intolerance, and by an absence of effective strategic urban planning that would provide non-confrontational outlets for all practitioners to play their parts in harmony – with each other as well as with society. Gustavo Giovannoni expressed very clearly the need for architects to have an integrated training: for them to be true *master builders*.

Architects have a key role to play in the conservation and sustainable development of historic cities, and there is an urgent need to address an inclusive profession (Figure 10.7).

Architectural conservation

Sir John Smith, founder of the Landmark Trust – which has restored a considerable number of historic properties across the United Kingdom and beyond – wrote in 1992 of the 'heresy which afflicts old buildings increasingly ... caused by the growth in the number of archaeologists. ... Good buildings ... are not archaeological sites, at least until they are ruins, and not

Creative continuity in architecture

Figure 10.8 Pulteney Bridge, Bath, England, was designed in neoclassical style by Robert Adam and completed in 1774. Lined with shops on both sides, it is supported on three elegant arches and features a central Venetian window.

Figure 10.9 Culzean Castle, Scotland, a mansion house designed by Robert Adam in a castellated, baronial manner externally; its neoclassical interiors include an oval-plan staircase (built 1777–90). Culzean Castle is now owned and administered as a visitor attraction by the National Trust for Scotland.

All creative work is derivative directly or indirectly of something, somewhere.

The Italian Renaissance recognized classical antiquity as a springboard for creative continuity. Today, references to historical styles in architecture are dismissed by some as *pastiche*, a term that is used in a derogatory sense as the antithesis of *contemporary*.

Was the Scots-born eighteenth-century architect Robert Adam (1728–92), equally adept in neoclassical, Gothic, picturesque and castle styles simply an architect of pastiche or a skilled and creative manipulator of form, construction and detail? (Figures 10.8 and 10.9.)

Was the Russian architect Alexander Schusev (1873–1949), practising in Moscow in the early-twentieth century in diverse styles – including Novgorodian Medieval, Art Nouveau, Russian and Tartar Revivalism, Neoclassicism and Constructivism – and whose works include the revivalist Kazan railway station (opened 1912) and the modernist Lenin Mausoleum in Red Square (completed 1930), a contemporary architect or a man of multiple pastiche?

Both Robert Adam and Alexander Schusev lived in places and during times in which an inclusive approach to creative continuity in architecture was recognised and respected.

It is a contrived debate, one that is essentially a product of intolerant twentieth-century modernism (Figures 10.10 and 10.11).

Figures 10.10 and 10.11 Pécs, Hungary, is noted for numerous small-scale contemporary interventions within its historic centre and for its many figurative and abstract sculptures in the public domain.

Contemporary architecture and the obsession with design

Contemporary building design is often perceived to substitute novelty for creativity in a self-conscious attempt to shock and attract attention.

Speaking in Edinburgh in September 2006, Francesco Bandarin, director of the UNESCO World Heritage Centre, spoke of a 'dangerous attitude, ... a growing gap between what architects want to do – and *maybe* their clients too – and what citizens want'. He referred to 'architecture seen as an object with a design, the artistic achievement of the architect, ... as something that demands to be seen', and of architects who are 'looking in the wrong direction: inwards to themselves, not outwards to the urban contexts in which they are building'.

Likening its appearance to a petrochemical factory, Bandarin identified the Georges-Pompidou Centre in Paris (see Figure 6.9 at page 123) as having initiated the present mould. He referred also to the Kunsthaus at Graz in Austria (completed in 2003, it is variously nicknamed the 'friendly alien' and the 'inflatable pigskin'; Peter Cook, architect) and the glittering shroud that envelopes the proposed extension to the neoclassical Marinsky Theatre, St Petersburg, Russia. He suggested that the best place for such buildings is at airports, well outside historic city centres.

On the positive side, I M Pei's pyramid in the forecourt of the Louvre, Paris, and Norman Foster's Carré d'Art that neighbours Maison Carrée, the Roman temple at Nîmes in the south of France, were applauded as 'brilliant solutions' for their self-effacing simplicity and elegance, ones that respond positively to the need for architectural continuity in historic cities (Figure 10.12).

The focus on design rather than on people and coordinated urban management has narrowed the appeal and the potential of the *Urban Renaissance* in Britain. As John Prescott, Deputy Prime Minister, colourfully put it: 'Architects think if they are not on board, it's a lousy design. All that's professional crap'.

Figure 10.12 Paris, France. The Pyramid in the forecourt of the Louvre (completed 1989; Ieoh Ming Pei, architect). The form of the pyramid is of course closely associated with structures from Ancient Egypt.

always even then . . . . Buildings are increasingly to be treated as documents, to be preserved for study, not as visual objects at all, and not for use'.

Fundamentalist attitudes to conservation charters and their like are commonplace. They do not communicate well to a generalist audience and frequently excite confrontation with building owners and users.

The philosophy and practical models of architectural conservation have a great deal to offer as contributions to sustainability. These contributions include safeguarding local distinctiveness; the reuse of buildings and the recycling of materials; the use of locally sourced building materials and craft skills; economical solutions based on an understanding of the environmental performance of historic structures; and the principle of minimum intervention.

Focusing on human and practical issues with which the non-academic, non-professional audience can identify, and relating these directly to the '3 Es' of sustainability and the '3 Rs' of non-renewable resource and waste management, underpins the response to Jukka Jokilehto's call for conservation to be redefined.

The opportunity

Just as architectural conservation has evolved from the eighteenth century, so has modern town planning from the nineteenth, and urban conservation from the twentieth. We have all moved on, and debates in future will increasingly focus on environmental issues under the umbrella concept of sustainability.

Academic criteria that address cultural significance from a narrow perspective, that are then not interpreted into policies that safeguard identity of place and local distinctiveness, and that fail to position conservation as a central activity in the public mind and within the construction professions and industry, neglect to respond effectively to the expectations of international agendas in either *conservation* or *sustainability*.

Civil society is enthused about *heritage*. It is extensively and progressively involved in the saving and restoration of individual buildings and building types. In the past it has successfully campaigned against and drawn a line under large-scale programmes of clearance and superimposed redevelopment in historic cities.

Today's challenges in the United Kingdom are principally longer term. The historic centres of many cities have three primary – if not exclusive – functions: employment and retail by day; entertainment by night. They remain largely deserted of resident populations, and sanitised of small-scale independent retailers, artisan-type workshops, and other traditional and localised activities.

The – as opposed to an – *Urban Renaissance* is a long way off. The interdisciplinary, cross-sector management skills are not in place. There are shortages of relevant skills in urban planning, architecture and conservation.

There is an absence of focus on people and an excess of focus on archaeological and design approaches to buildings and cities as physical objects.

The opportunity that now arises may be summarised as follows:

- to redefine *conservation* to make it relevant to *sustainability*;
- to reverse the anti-urban legacy and redefine the city to make it relevant to citizens as a place in which to live as well as to work, shop and play;
- to adopt a resource management approach in which material and cultural resource values act as mutually supportive partners;
- to progressively reorder existing, historic cities for a sustainable future, recognising and safeguarding all of the tangible and intangible cultural values that are associated with them; and
- to address the urban conservation challenge holistically and position *conservation* as a determining factor in *sustainable development*.

In the United Kingdom there is much lost ground to be made up. Elsewhere in the world, especially in emerging and developing economies, there is an urgent need to avoid replicating non-sustainable theoretical and practical models.

Sources

Chapter 1

Architectural conservation: beginnings and evolution

Jokilehto, Jukka, *A History of Architectural Conservation*, Elsevier Butterworth-Heinemann, Oxford (2004); first published, Butterworth-Heinemann, Oxford (1999).

Earl, John, *Building Conservation Philosophy*, Donhead Publishing, Donhead St Mary (third edn, 2003); first published, College of Estate Management, Reading (1996); second edn (1997).

Gilpin, William, *Essay on Prints*, J. Robson, London (1768).

Miers, Mary, 'To restore or not to restore?', in *Country Life*, IPC, London (22 September 2005).

UNESCO, *Convention Concerning the Protection of the World Cultural and Natural Heritage* (the *World Heritage Convention*), UNESCO, Paris (1972).

Pevsner, Nikolaus, 'Scrape and anti-scrape', in: Fawcett, Jane (ed.), *The Future of the Past*, Thames and Hudson, London (1976).

Fawcett, Jane, 'A restoration tragedy: cathedrals in the eighteenth and nineteenth centuries', in: Fawcett, Jane (ed.), *The Future of the Past*, Thames and Hudson, London (1976).

Viollet-le-Duc, Eugène Emmanuel, *Dictionnaire raisonné de l'architecture française du XIe au XVIe siècle*, vol. VIII, Paris (1866).

Pound, Christopher, *Genius of Bath: The City and its Landscape*, Millstream Books, Bath (1986).

Ferguson, Adam, *The Sack of Bath*, Compton Russell, Salisbury (1973).

Buchanan, Colin and Partners, *Bath: A Planning and Transport Study*, Colin Buchanan, London (1965).

Richards, James (ed.), *European Heritage*, Phoebus Publishing, London (Issues One to Five, 1975).

Strong, Roy et al., *The Destruction of the Country House: 1875 to 1975*, Thames and Hudson, London (1974).

Department of the Environment, *What is our Heritage? – United Kingdom Achievements for European Architectural Heritage Year 1975*, HMSO, London (1975).

SAVE Britain's Heritage, 'The SAVE Report', in *Architects' Journal*, Architectural Press, London (17/24 December 1975).

The language of architectural conservation

UNESCO, 'Heritage: A Gift From The Past To The Future', part of the mission statement of the World Heritage Centre, UNESCO, Paris (May 2000).

Definitions of *heritage* on the web.
Lowenthal, David, *The Past is a Foreign Country*, Cambridge University Press, Cambridge (1985).
Lowenthal, David, *The Heritage Crusade and the Spoils of History*, Cambridge University Press, Cambridge (1998).
Fladmark, Magnus (ed.), *Cultural Tourism*, Donhead Publishing, Donhead St Mary (1994).
Arnold, John et al. (eds), *History and Heritage: Consuming the Past in Contemporary Culture*, Donhead Publishing, Donhead St Mary (1998).
Feilden, Bernard, *Conservation of Historic Buildings*, Architectural Press, London (third edn, 2003); first published, Architectural Press, London (1982).
Feilden, Bernard and Jokilehto, Jukka, *Management Guidelines for World Cultural Heritage Sites*, ICCROM, Rome (second edn, 1998); first published (1993).

Conservation charters

Choay, Françoise (ed.), *La Conférence d'Athènes sur la conservation artistique et historique des monuments*, Editions de l'Imprimeur, Paris (2002).
Rodwell, Dennis, 'Dubrovnik, Pearl of the Adriatic', in *World Heritage Review*, UNESCO, Paris, and Ediciones San Marcos, Madrid (December 2004).
Morris, William, *Manifesto of the Society for the Protection of Ancient Buildings* (the *SPAB Manifesto*), SPAB, London (1877).
The Athens Charter for the Restoration of Historic Monuments (the *Athens Charter*), adopted at the First International Congress of Architects and Technicians of Historic Monuments, Athens (1931).
La Charte d'Athènes, adopted at the Fourth CIAM (Congrès Internationaux d'Architecture Moderne) Congress, Paris (1933).
International Charter for the Conservation and Restoration of Monuments and Sites (the *Venice Charter*), adopted at the Second International Congress of Architects and Technicians of Historic Monuments, Venice (1964).
Council of Europe, *European Charter of the Architectural Heritage* (the *European Charter*), Council of Europe (1975).
The Declaration of Amsterdam, adopted at the Congress on the European Architectural Heritage, Amsterdam (1975).
ICOMOS, *The Charter for the Conservation of Historic Towns and Urban Areas* (the *Washington Charter*), adopted at the ICOMOS General Assembly, Washington (1987).
ICOMOS-Australia, *The Australia ICOMOS Charter for Places of Cultural Significance* (the *Burra Charter*), ICOMOS-Australia (revision, November 1999); first published, ICOMOS-Australia (1979).
Kerr, James Semple, *Sydney Opera House: An Interim Plan for the Conservation of the Sydney Opera House and its Site*, Sydney Opera House Trust, Sydney (1993).

Urban conservation: museological beginnings

Sorlin, François, 'Paris', in: Matthew, Robert (ed.), *The Conservation of Georgian Edinburgh*, Edinburgh University Press, Edinburgh (1972).
Boucly, Yves (ed.), *Administration: Revue d'Information*, L'Association du corps préfectoral et des hauts fonctionnaires du Ministère de l'Intérieur, Paris (June 1971).
Hartmann, Véronique (ed.), *Centres et Quartier Anciens*, Les Monuments Historiques de la France, Paris (undated, *c.*1976).
Rodwell, Dennis, 'The French connection: The significance for Edinburgh of conservation policies in the Marais, Paris', in: Harrison, Patrick (ed.), *Civilising the City: Quality or Chaos in Historic Towns*, Nic Allen, Edinburgh (1990).

Rodwell, Dennis, 'Conservation on the Continent: A Report on the Protection of Historic Buildings and Urban Areas in France, Italy, Austria and Germany', unpublished, Cambridge (1972).
Kennet, Lord, 'Notes of a Visit to *Secteurs Sauvegardés* in France: June 1966', unpublished paper (1966).
Rodwell, Dennis, *The Conservation of Monuments in the 'Ancient Plovdiv Reserve', Plovdiv, Bulgaria* (mission Report), Division of Cultural Heritage, UNESCO, Paris (2002).
Rodwell, Dennis, 'Approaches to Urban Conservation in Central and Eastern Europe' in *Journal of Architectural Conservation*, Donhead Publishing, Donhead St Mary (July 2003).

Townscape: the concept

Cullen, Gordon, *Townscape*, Architectural Press, London (1961).
Lynch, Kevin, *The Image of the City*, MIT Press, Cambridge, Massachusetts (1960).

Chapter 2

The pre-industrial city

Kostof, Spiro, *The City Shaped*, Thames and Hudson, London (1991).
Kostof, Spiro, *The City Assembled*, Thames and Hudson, London (1992).
Rodwell, Dennis, 'Making Cities' (book review), in *RSA Journal*, Royal Society of Arts, London, (July 1993).
Mumford, Lewis, *The City in History*, Secker and Warburg, London (1961).
Bell, Colin and Rose, *City Fathers: The Early History of Town Planning in Britain*, Penguin Books, Harmondsworth (1972); first published by Barrie and Rockliff, Cresset Press (1969).
Rodwell, Dennis, 'Dubrovnik, Pearl of the Adriatic', in *World Heritage Review*, UNESCO, Paris, and Ediciones San Marcos, Madrid (December 2004).
Rasmussen, Steen Eiler, *Towns and Buildings*, Liverpool University Press, Liverpool (1951); first published in Denmark (1949).
Rosenau, Helen, *The Ideal City: Its Architectural Evolution in Europe*, Methuen, London (third edn, 1983); first published as: *The Ideal City in its Architectural Evolution*, Routledge and Kegan Paul, London (1959); second edn, Studio Vista, London (1974).
Kowalczyk, Jerzy, *Zamość: Città Ideale in Polonia*, Narodowa Instytucja Kultury, Warsaw (1994).
Lipiec, Wiesław, *Zamość*, Parol, Krakow (1997).

The mainstream of modern town planning

Briggs, Asa, *Victorian Cities*, Odhams, London (1963).
Ackroyd, Peter (ed.), *Dickens' London: An Imaginative Vision*, Book Club Associates, Swindon (1987); first published, Headline, London (1987).
Canning, John (ed.), *The Illustrated Mayhew's London*, Book Club Associates, Swindon (1986); first published, Weidenfeld and Nicolson, London (1986).
Benevolo, Leonardo, *The Origins of Modern Town Planning*, Routledge and Kegan Paul, London (1957); first published as: *Le Origini dell'Urbanistica Moderna*, Laterza, Bari (1963).
Choay, Françoise, *L'urbanisme, Utopies et Réalités: Une Anthologie*, Seuil, Paris (1965).
Charlton, Christopher (ed.), *The Derwent Valley Mills and their Communities*, Derwent Valley Mills Partnership, Matlock, (2001).

Rodwell, Dennis, 'Industrial World Heritage sites in the United Kingdom', in *World Heritage Review*, UNESCO, Paris, and Ediciones San Marcos, Madrid (December 2002).

Alvès, Gilles, *et al* (ed.), *Patrimoine industriel: cinquante sites en France*, Editions du patrimoine, Paris (1997).

Howard, Ebenezer, *Garden Cities of To-morrow*, edited with a preface by F. J. Osborn and an introductory essay by Lewis Mumford, Faber and Faber, London (1946); first published as *To-Morrow: A Peaceful Path to Real Reform*, London (1898); second edn published as *Garden Cities of To-morrow*, Swan Sonnenschein, London (1902).

La Charte d'Athènes, adopted at the Fourth CIAM (Congrès Internationaux d'Architecture Moderne) Congress, Paris (1933).

Le Corbusier, *La Charte d'Athènes*, Editions de Minuit, Paris, (1957).

Le Corbusier, *The City of To-morrow and its Planning*, edited with a preface by Le Corbusier, Architectural Press, London (1947); first published as *Urbanisme*, Editions Crés, Paris (1924); first English edition published by John Rodker, London (1929).

Maulini, Marcel, *Comprendre Ronchamp*, Chazelle, Dole, Jura (1964).

Paquot, Thierry, 'Le XXe siècle: de la ville à l'urbain: Chronique urbanistique et architecturale de 1900 à 1999', in *Urbanisme*, Sarl, Paris (November/December 1999).

Sharp, Thomas, *Oxford Replanned*, Architectural Press, London (1948).

Johnson-Marshall, Percy, *Rebuilding Cities*, Edinburgh University Press, Edinburgh (1966).

Tetlow, John and Goss, Anthony, *Homes, Towns and Traffic*, Faber and Faber, London (second edn, 1968); first published, Faber and Faber, London (1965).

Holliday, John (ed.), *City Centre Redevelopment*, Charles Knight, London (1973).

Cherry, Gordon, *Cities and Plans: The Shaping of Britain in the Nineteenth and Twentieth Centuries*, Edward Arnold, London (1988).

Middleton, Michael, *Cities in Transition*, Michael Joseph, London (1991).

Cullingworth, Barry and Nadin, Vincent, *Town & Country Planning in the UK*, Routledge, London (twelfth edn, 1997); first published (1964).

Alternative visions

Meller, Helen, *Patrick Geddes: Social Evolutionist and City Planner*, Routledge, London (1990).

Boardman, Philip, *The Worlds of Patrick Geddes: Biologist, Town Planner, Re-educator, Peace-Warrior*, Routledge and Kegan Paul, London (1978).

Welter, Volker, *Biopolis: Patrick Geddes and the City of Life*, MIT Press, Cambridge, Massachusetts (2002).

Geddes, Patrick, *Cities in Evolution*, with an introduction by Percy Johnson-Marshall, Ernest Benn, London (1968); first published by Williams and Norgate, London (1915).

Welter, Volker, *Collecting Cities: Images from Patrick Geddes' Cities and Town Planning Exhibition*, Collins Gallery, Glasgow (1999).

Reid, Lindsay, 'Old Town opens bright new future', in *Evening News*, Scotsman Publications, Edinburgh (29 April 1985).

Burns, Wilfred, *New Towns for Old: The Technique of Urban Renewal*, HMSO, London (1963). Page 93 contains the following: 'One result of slum clearance is that a considerable movement of people takes place over long distances, with devastating effect on the social grouping built up over the years. But, one might argue, this is a good thing when we are dealing with people who have no initiative or civic pride. The task, surely, is to break up such groupings even though the people seem to be satisfied with their miserable environment and seem to enjoy an extrovert social life in their own locality.'

Young, Michael and Willmott, Peter, *Family and Kinship in East London*, Penguin, London (revised edn with new introduction, 1986); first edn, Routledge and Kegan Paul, London (1957).

Jacobs, Jane, *The Death and Life of Great American Cities: The Failure of Town Planning*, Random House, New York (1961).

Coleman, Alice, *Utopia on Trial: Vision and Reality in Planned Housing*, Hilary Shipman, London (1985).

Christine, F., *Moi, Christiane F., 13 ans, droguée, prostituée*, Mercure de France, Saint-Amande (1989); first published as *Christine F. Wir Kinder von Bahnhof Zoo*, Stern-Magazin, Hamburg (1978). The identity of 'F.' is not disclosed.

U'ren, Graham (ed.), 'Sir Patrick Geddes (1854–1932)', in *Scottish Planner*, Royal Town Planning Institute in Scotland (February 2004).

Giovannoni, Gustavo, *L'urbanisme face aux villes anciennes*, with an introduction by Françoise Choay, Seuil, Paris (1998); first published as *Vecchie città ed edilizia nuova*, UTET Libreria, Rome (1931); second edn, CittàStudi Edizione, Rome (1995). Giovannoni first set out the principal elements of his thesis in a set of papers that were published in 1913 under the title *Vecchie città ed edilizia nuova: il quartiere del Rinascimento in Roma*.

Jokilehto, Jukka, *A History of Architectural Conservation*, Elsevier Butterworth-Heinemann, Oxford (2004); first published Butterworth-Heinemann, Oxford (1999).

Giovannoni, Gustavo, 'La restauration des monuments en Italie' and 'Les moyens modernes de construction appliqués à la restauration des monuments', in: Choay, Françoise (ed.), *La Conférence d'Athènes sur la conservation artistique et historique des monuments*, Editions de l'Imprimeur, Paris (2002).

McKean, John, *Giancarlo De Carlo: Layered Places*, Axel Menges, Stuttgart (2004).

Summerson, John, *Georgian London*, Yale University Press, New Haven and London (2003); this revised edn was edited posthumously by Howard Colvin; first edn, Pleiades Books, London (1945); second edn, Penguin Books, Harmondsworth (1962).

Rasmussen, Steen Eiler, *London: The Unique City*, Jonathan Cape, London (revised edn, 1937); first published in Denmark (1934); abridged edn, Penguin Books, Harmondsworth (1960).

Urban conservation: mainstream beginnings in the United Kingdom

Buchanan, Colin, *Traffic in Towns*, HMSO, London (1963).

Buchanan, Colin, *Traffic in Towns: The Specially Shortened Edition of the Buchanan Report*, Penguin Books, Harmondsworth (1964).

Richards, James (ed.), 'Dropping the Pilot?', in *Architectural Review*, Architectural Press, London (December 1970).

Buchanan, Colin and Partners, *Bath: A Planning and Transport Study*, Colin Buchanan, London (1965).

Buchanan, Colin and Partners, *Bath: A Study in Conservation*, HMSO, London (1968).

Rodwell, Dennis, 'City of Bath: A Masterpiece of Town Planning', in *World Heritage Review*, UNESCO, Paris, and Ediciones San Marcos, Madrid (October 2005).

Insall, Donald and Associates, *Chester: A Study in Conservation*, HMSO, London (1968).

Insall, Donald, *Conservation in Action: Chester's Bridgegate*, HMSO, London (1982).

Insall, Donald, *Conservation in Chester*, Chester City Council, Chester (1986).

Burrows, G., *Chichester: A Study in Conservation*, HMSO, London (1968).

Esher, Viscount, *York: A Study in Conservation*, HMSO, London (1968).

Cummin, David (ed.), *York 2000: People in Protest*, York 2000, York (1973).

Simpson, James, 'Whither conservation?', in: Harrison, Patrick (ed.), *Civilising the City: Quality or Chaos in Historic Towns*, Nic Allen, Edinburgh (1990).

Ward, Pamela (ed.), *Conservation and Development in Historic Towns and Cities*, Oriel Press, Newcastle-upon-Tyne (1968).

The character of towns

Worskett, Roy, *The Character of Towns: An Approach to Conservation*, Architectural Press, London (1969).

Worskett, Roy, 'Great Britain: Progress in Conservation', in: Cantacuzino, Sherban (ed.), *Architectural Conservation in Europe*, Architectural Press, London (1975).

Chapter 3

Sustainability: beginnings and evolution

Dresner, Simon, *The Principles of Sustainability*, Earthscan Publications, London (2002).

Malthus, Thomas, *An Essay on the Principle of Population*, Oxford University Press, Oxford (1999); first published anonymously (1798); second revised edn, Johnson, London (1803); revised again several times during Malthus' life-time.

Ehrlich, Paul, *The Population Bomb*, Ballantine, New York (1968).

Shelley, Mary, *The Last Man*, Wildside Press, Doylestown, Pennsylvania (2004); first published Henry Colburn, London (1826). Mary (Wollstonecraft) Shelley, second wife of the poet Percy Bysshe Shelley, is best known as the author of *Frankenstein*, first published in 1818.

Huxley, Aldous, *Brave New World*, Chatto and Windus, London (1932).

Huxley, Aldous, *Brave New World Revisited*, Harper, New York (1958).

Orwell, George, *Nineteen Eighty-Four*, Secker and Warburg, London (1949).

Scott, Peter, *The Eye of the Wind*, Hodder and Stoughton, London (1961).

Botting, Douglas, *Gerald Durrell: The Authorized Autobiography*, Carroll & Graf, New York (1999). Gerald Durrell's first book, *The Overloaded Ark*, was published by Faber & Faber, London, in 1953.

Attenborough, David, *Life on Air: Memoirs of a Broadcaster*, BBC, London (2002).

Carson, Rachel, *Silent Spring*, Houghton Mifflin, Boston (1962).

Ward, Barbara and Dubos, René, *Only One Earth: The Care and Maintenance of a Small Planet*, Norton, New York (1972).

Meadows, Donella et al., *The Limits to Growth*, Universe Books, New York (1972).

Meadows, Dennis et al., *Dynamics of Growth in a Finite World*, Wright-Allen Press, Cambridge, Massachusetts (1974).

Meadows, Donella et al., *Beyond the Limits*, Chelsea Green, Vermont (1992).

Meadows, Donella et al., *Limits to Growth: The 30-Year Update*, Chelsea Green, Vermont (2004).

World Commission on Environment and Development, *Our Common Future* (known as the *Brundtland Report*), Oxford University Press, Oxford (1987).

Elliott, Jennifer, *An Introduction to Sustainable Development*, Routledge, London (second edn, 1999); first published, Routledge, London (1994).

Rodwell, Dennis, 'Balance and involvement in a world heritage city', in *Context*, Institute of Historic Building Conservation (December 2000).

The language of sustainability

Definitions of *sustainable development*, *sustainability* and *sustainable* on the web.

Chase, Martin, 'Are our town centres sustainable?', in *RSocietyAJournal*, Royal Society of Arts, London (July 1995).

Rodwell, Dennis, 'Managing urban world heritage cities – the way ahead', in *Context*, Institute of Historic Building Conservation (November 2002).

Feilden, Bernard, 'Conservation – Is There No Limit? – A Review', in *Journal of Architectural Conservation*, Donhead, London (March 1995).

Urban conservation: strategic beginnings at the metropolitan scale

Hall, Peter, *The World Cities*, World University Library, Weidenfeld and Nicolson, London (1966).
Ministry of Housing and Local Government, *The South East Study: 1961–1981*, HMSO, London (1964).
Hall, Peter, *London 2000*, Faber and Faber, London (second edn, 1969); first published, Faber and Faber, London (1963).
Burtenshaw, David et al., *The European City: A Western Perspective*, David Fulton, London (1991).
Bell, Colin and Rose, *City Fathers: The Early History of Town Planning in Britain*, Penguin Books, Harmondsworth (1972); first published by Barrie and Rockliff, Cresset Press (1969).
Rodwell, Dennis, 'The French connection: the significance for Edinburgh of conservation policies in the Marais, Paris', in: Harrison, Patrick (ed.), *Civilising the City: Quality or Chaos in Historic Towns*, Nic Allen, Edinburgh (1990).
Gravier, Jean-François, *Paris et le Désert Français*, Le Portulan, Paris (1947).
Ambassade de la France, *France: Town and Country Environmental Planning*, Service de Presse et de l'Information, New York (1965).
Marchand, Bernard, *Paris, histoire d'une ville (XIXe –XXe siècle)*, Seuil, Paris (1993).
Lacaze, Jean-Paul, *Paris: urbanisme d'état et destin d'une ville*, Flammarion, Paris (1994).
Chieng, Diana Chan, *Projets Urbains en France*, Le Moniteur, Paris (2002).
Arondel, Mathilde, *Chronologie de la politique urbaine: 1945–2000*, ANAH, Paris (2001).
Kimmel, Alain (ed.), *Les Villes Nouvelles en Ile-de France*, Echos, Paris (1988).
Jodido, Philip, *Grands Travaux*, Connaissance des Arts, Paris (1992).
Girard, Joël, 'Le Triomphe de l'Arche', in *Atlas Air France*, Editions Atlas, Paris (March 1990).
Rodwell, Dennis, 'The re-shaping of Paris', unpublished paper (1998).

Chapter 4

UNESCO

UNESCO, *Convention Concerning the Protection of the World Cultural and Natural Heritage* (the *World Heritage Convention*), UNESCO, Paris (1972).
UNESCO, *Operational Guidelines for the Implementation of the World Heritage Convention*, UNESCO, Paris (latest revision, February 2005); first published, UNESCO, Paris (1977).
UNESCO, 'Mission Statement' and associated set of papers, the World Heritage Centre, UNESCO, Paris (May 2000).
UNESCO, *Tell me about UNESCO*, UNESCO, Paris (2002).
UNESCO, *Tell me about World Heritage*, UNESCO, Paris (2002).
UNESCO, *Recommendation Concerning the Protection, at National Level, of the Cultural and Natural Heritage*, UNESCO, Paris (1972).
Pressouyre, Léon, *The World Heritage Convention, twenty years later*, UNESCO, Paris (1996).
Batisse, Michel and Bolla, Gerard, *The Invention of 'World Heritage'*, Association of Former UNESCO Staff Members, Paris (2005).
Rodwell, Dennis, 'The World Heritage Convention and the Exemplary Management of Complex Heritage Sites', in *Journal of Architectural Conservation*, Donhead Publishing, Donhead St Mary (November 2002).

Rodwell, Dennis, 'Industrial World Heritage sites in the United Kingdom', in *World Heritage Review*, UNESCO, Paris, and Ediciones San Marcos, Madrid (December 2002).

Rodwell, Dennis, 'City of Bath: A Masterpiece of Town Planning', in *World Heritage Review*, UNESCO, Paris, and Ediciones San Marcos, Madrid (October 2005).

Cunliffe, Barry, *The Roman Baths at Bath*, Bath Archaeological Trust, Bath (1993).

Rodwell, Dennis, 'Philosophy and practice' (published as a letter), in *Context*, Institute of Historic Building Conservation (July 2003).

ICOMOS-German Democratic Republic, *The Declaration of Dresden*, ICOMOS-GDR, Dresden (1982).

ICOMOS, 'Advisory body evaluation: The Historic Center of Warsaw, Poland', ICOMOS, Paris (1980).

ICOMOS, 'Advisory body evaluation: Rila Monastery, Bulgaria', ICOMOS, Paris (1983).

ICOMOS, 'Advisory body evaluation: Urnes Stave Church, Norway', ICOMOS, Paris (1978).

ICOMOS, 'Advisory body evaluation: The Wooden Churches of Maramures, Romania', ICOMOS, Paris (1999).

ICOMOS, 'Advisory body evaluation: Bryggen quarter, Bergen, Norway', ICOMOS, Paris (1978).

ICOMOS, 'Advisory body evaluation: Old Rauma, Finland', ICOMOS, Paris (1990).

Lemaire, Raymond and Stovel, Herb (eds), *Nara Document on Authenticity*, Nara, Japan (1994).

ICOMOS, 'Advisory body evaluation: The Historic Fortified Town of Carcassonne, France', ICOMOS, Paris (1997).

Boulting, Nikolaus, 'The law's delays: conservation legislation in the British Isles', in: Fawcett, Jane (ed.), *The Future of the Past*, Thames and Hudson, London (1976).

ICOMOS, 'Advisory body evaluation: Durham Cathedral and Castle, England', ICOMOS, Paris (1986).

Bortolotto, Chiara, 'From the monumental to the living heritage: a shift in perspective', in: Carman, John and White, Roger (eds), *World Heritage: Global Challenges and Local Solutions*, British Archaeological Reports, International Series, Archaeopress, Oxford (forthcoming).

UNESCO, *Convention for the Safeguarding of the Intangible Cultural Heritage*, UNESCO, Paris (2003).

ICOMOS, 'Advisory body evaluation: The Blaenavon Industrial Landscape, United Kingdom', ICOMOS, Paris (2000).

Rodwell, Dennis, 'Community involvement in the regeneration of a historic city centre', in: Carman, John and White, Roger (eds), *World Heritage: Global Challenges and Local Solutions*, British Archaeological Reports, International Series, Archaeopress, Oxford (forthcoming).

Rodwell, Dennis, 'Managing Historic Cities: the Management Plans for the Bath and Edinburgh World Heritage Sites', in *Journal of Architectural Conservation*, Donhead Publishing, Donhead St Mary (July 2006).

Rodwell, Dennis, 'The values of world heritage', in *Context*, Institute of Historic Building Conservation (July 2006).

Urban conservation: international cooperation

Rowell, Stephen et al., *A History of Lithuania*, Inter Se, Vilnius (2002).

Žukas, Saulius, *Vilnius: The City and its History*, Inter Se, Vilnius (2002).

Rodwell, Dennis, 'Approaches to Urban Conservation in Central and Eastern Europe', in *Journal of Architectural Conservation*, Donhead Publishing, Donhead St Mary (July 2003).

Vilnius City Municipality, *Vilnius City Strategic Plan: 2002–2011*, Vilnius City Municipality, Vilnius (2000).

Vilnius Old Town Renewal Agency, *Vilnius Old Town Revitalisation Strategy Implementation: Co-operation, Results, Vision*, OTRA, Vilnius (2001).
Vilnius City Municipality, *Vilnius City Official Plan: Summary of Main Statements*, Vilnius City Municipality, Vilnius (2000).
Raugelienė, Jūratė, 'The Revitalisation of Vilnius Old Town', in: UNESCO, *Management of Private Property in the Historic City-Centres of the European Countries-in-Transition*, Division of Cultural Heritage, UNESCO, Paris (2002).
Vilnius Old Town Renewal Agency with Thomson, Kirsteen, *Conservation Guidelines*, OTRA, Vilnius (2002).
Bardauskienė, Dalia et al. (eds), *Vilnius Old Town Revitalisation: 1998–2003*, Jsc Ikstrys, Vilnius (2003).
Rutkauskas, Gediminas et al., *Newsletter*, ICCROM Integrated Territorial and Urban Conservation Program for North-East Europe, Vilnius (May 2003).
Rodwell, Dennis, 'Management of historic cities in the European Countries-in-Transition: experience and problems', paper delivered at the conference *Continuity of Urban Development in Historic Cities*, Vilnius, Lithuania (June 2003).
Rodwell, Dennis, *The Revitalisation of World Heritage Cities in Central and Eastern Europe* (mission report), World Heritage Centre, UNESCO, Paris (May 1999).
Rodwell, Dennis, *The Revitalisation of Vilnius Old Town, Lithuania* (mission report), World Heritage Centre, UNESCO, Paris (June 1999).
Rodwell, Dennis, 'Dubrovnik, Pearl of the Adriatic', in *World Heritage Review*, UNESCO, Paris, and Ediciones San Marcos, Madrid (December 2004).
Harris, Robin, *Dubrovnik: A History*, Saqi, London (2003).
Foretić, Miljenko, *Dubrovnik in War*, Matica Hrvatska, Dubrovnik (eleventh edn, 2002); first published Dubrovnik (1993).
Letunić, Božo, *The Restoration of Dubrovnik: 1979–89*, Institute for the Restoration of Dubrovnik, Dubrovnik (1990).
Institute for the Restoration of Dubrovnik, *Die Erneuerung Dubrovniks*, Institute for the Restoration of Dubrovnik, Dubrovnik (undated, c1996).
Rodwell, Dennis, *The ARCH Foundation and the Conservation of Monuments on the Island of Lopud, Dalmatia, Croatia* (mission report), Division of Cultural Heritage, UNESCO, Paris (2003).

Chapter 5

Established protective system

Rodwell, Dennis, 'Conservation legislation', in: Cantacuzino, Sherban (ed.), *Architectural Conservation in Europe*, Architectural Press, London (1975).
Rodwell, Dennis, 'The Minto House debacle', in *Context*, Association of Conservation Officers (December 1992).

The involvement of civil society, technical and policy guidance

Watters, Diane and Glendinning, Miles, 'Size matters', in *Scotland in Trust*, The National Trust for Scotland, Edinburgh (Spring 2006).
Watters, Diane and Glendinning, Miles, *Little Houses: The National Trust for Scotland's Improvement Scheme for Small Houses*, The Royal Commission on the Ancient and Historical Monuments of Scotland and The National Trust for Scotland, Edinburgh (2006).

Rodwell, Dennis, 'A house in the country', in *Scottish Field*, Holmes McDougall, Glasgow (June 1985).

Greysmith, Brenda, 'Little house on the Green', in *Traditional Homes*, Benn, St Albans (April 1987).

AHF, *Annual Review 2004–05*, The Architectural Heritage Fund, London (2005). This lists the total of 171 building preservation trusts that appear on the AHF's register.

Powys, A.R., *The Repair of Ancient Buildings*, Society for the Protection of Ancient Buildings, London (reprinted 1981); first published, Dent, London (1929).

Insall, Donald, *The Care of Old Buildings Today: A Practical Guide*, Architectural Press, London (1972).

Feilden, Bernard, *Conservation of Historic Buildings*, Architectural Press, London (third edn, 2003); first published, Architectural Press, London (1982).

Fitch, James Marston, *Historic Preservation: Curatorial Management of the Built World*, University Press of Virginia, Charlottesville (1990); first published, McGraw Hill, New York (1982).

Ashurst, John and Ashurst, Nicola, *Practical Building Conservation: English Heritage Technical Handbook*, Gower Technical Press, London (Volumes 1 to 5, 1988).

Department of the Environment et al., *New Life for Old Buildings*, HMSO, London (1971).

Department of the Environment et al., *New Life for Historic Areas*, HMSO, London (1972).

Cantacuzino, Sherban (ed.), 'New Uses for Old Buildings', in *Architectural Review*, Architectural Press, London (May 1972).

Cormack, Patrick, *Heritage in Danger*, New English Library, London (1976).

Andreae, Sophie and Binney, Marcus, *Tomorrow's Ruins? Country Houses at Risk*, SAVE Britain's Heritage, London (1978).

Binney, Marcus et al., *Satanic Mills*, SAVE Britain's Heritage, London (undated, c.1978).

Binney, Marcus et al., *Lost Houses of Scotland*, SAVE Britain's Heritage, London (1980).

Andreae, Sophie et al., *Silent Mansions: More Country Houses at Risk*, SAVE Britain's Heritage, London (1981).

Binney, Marcus and Martin, Kit, *The Country House: To Be or Not To Be*, SAVE Britain's Heritage, London (undated, c.1982).

Binney, Marcus, *Our Vanishing Heritage*, Arlington Books, London (1984).

Wolton, Julia, *Endangered Domains*, SAVE Britain's Heritage, London (1985).

Pearce, David, *Conservation Today*, Routledge, London (1989).

Dean, Marcus and Miers, Mary, *Scotland's Endangered Houses*, SAVE Britain's Heritage, London (1990).

Binney, Marcus and Watson-Smyth, Marianne, *The SAVE Britain's Heritage Action Guide*, SAVE Britain's Heritage, London (1991).

Fladmar, Magnus (ed.), *Heritage Conservation, Interpretation and Enterprise*, Donhead Publishing, Donhead St Mary (1993).

Department of the Environment and Department of National Heritage, *Planning Policy Guidance: Planning and the Historic Environment* (known as PPG 15), HMSO, London (1994).

Department of the Environment, *Planning Policy Guidance: Archaeology and Planning* (known as PPG 16), HMSO, London (1990).

English Heritage, *Development in the Historic Environment: An English Heritage Guide to Policy, Procedure, and Good Practice*, English Heritage, London (1995).

English Heritage, *Conservation Area Practice: English Heritage Guidance on the Management of Conservation Areas*, English Heritage, London (1995).

English Heritage, *Sustaining the Historic Environment: New Perspectives on the Future: An English Heritage Discussion Document*, English Heritage, London (1997).

English Heritage, *Conservation Area Appraisals: Defining the Special Architectural or Historic Interest of Conservation Areas*, English Heritage, London (1997).

English Heritage, *Enabling Development and the Conservation of Heritage Assets: Policy Statement; Practical Guide to Assessment*, English Heritage, London (2001).

English Heritage and the Commission for Architecture and the Built Environment, *Building in Context: New Development in Historic Areas*, English Heritage/CABE, London (2001).

Clark, Kate, *Informed Conservation: Understanding Historic Buildings and their Landscapes for Conservation*, English Heritage, London (2001).

Rodwell, Dennis, 'Informed Conservation' (book review), *Journal of Architectural Conservation*, Donhead Publishing, Donhead St Mary (July 2002).

English Heritage, *Guidance on the Management of Conservation Areas*, English Heritage, London (2006).

English Heritage, *Conservation Principles for the Sustainable Management of the Historic Environment: First Stage Consultation*, English Heritage, London (2006).

Rodwell, Dennis, *Britannia Music Hall: Feasibility Study*, Dennis Rodwell Architects, Melrose (1993).

Maloney, Paul, 'Curtain call', in *Scottish Field*, Caledonian Magazine, East Kilbride (April 1994).

Wilkinson, Philip, *Restoration: Discovering Britain's Hidden Architectural Treasures*, Headline, London (2003).

Wilkinson, Philip, *Restoration: The Story Continues*, English Heritage, Swindon (2004).

Failures within the established protective system

Civic Amenities Act 1967, HMSO, London (1967). This Act applied to England, Wales and Scotland. The definition of conservation areas in this and subsequent statutory instruments is: 'areas of special architectural or historic interest the character or appearance of which it is desirable to preserve or enhance'. As the word *or* is not mutually exclusive, has been the subject of legal argument, and appears affected outside formal texts, the word *and* is substituted throughout this book.

Charlton, Christopher (ed.), *The Derwent Valley Mills and their Communities*, Derwent Valley Mills Partnership, Matlock, (2001).

Rodwell, Dennis, *Darley Abbey, Derby: Historical and Architectural Notes on Surviving Evans Buildings*, Derby City Council, Derby (2001).

Rodwell, Dennis, 'Darley Abbey, Derby: the settlement – historic door and window detailing' (departmental record), Derby City Council, Derby (2002).

Rodwell, Dennis, *Darley Abbey & Park to Derby Silk Mill: Part of the Derwent Valley Mills World Heritage Site* (visitor leaflet), Derby City Council, Derby (2002).

Rodwell, Dennis, *Darley Abbey Conservation Area Appraisal*, Derby City Council, Derby (revised for publication, 2003; publication forthcoming).

English Historic Towns Forum, *Townscape in Trouble: Conservation Areas – The Case for Change*, English Historic Towns Forum, Bath (1992).

English Historic Towns Forum, *Conservation Area Management – A Practical Guide*, English Historic Towns Forum, Bath (1998).

Department of the Environment, Transport and the Regions, and Department for Culture, Media and Sport, 'Joint DETR/DCMS Consultation Paper on the Impact of the Shimizu Judgement'. DETR and DCMS, London (19 June 2000).

Rodwell, Dennis, *The Revitalisation of World Heritage Cities in Central and Eastern Europe* (mission report), World Heritage Centre, UNESCO, Paris (May 1999).

Review of policies relating to the historic environment in England

English Heritage, 'Government's Review of Policies Relating to the Historic Environment: An Invitation to Participate' English Heritage, London (1 February 2000). Also its appendix: letter dated 31 January 2000 from Alan Howarth MP, Minister for the Arts, to Sir Jocelyn Stevens, Chairman of English Heritage, formally commissioning the review.

Rodwell, Dennis, letter to English Heritage (17 March 2000).

English Heritage, 'Review of Policies Relating to the Historic Environment: Consulation' English Heritage, London (June 2000). This set of documents comprises five discussion papers and a sixth document posing a list of questions. The full definition of *historic environment*, which appears in Discussion Document 1, reads as follows:

> The historic environment is all the physical evidence for past human activity, and its associations, that people can see, understand and feel in the present world.
> ☐ It is the habitat that the human race has created through conflict and cooperation over thousands of years, the product of human interaction with nature
> ☐ It is all round us as part of everyday experience and life, and it is therefore dynamic and continually subject to change
> At one level, it is made up entirely of places such as towns or villages, coast or hills, and things such as buildings, buried sites and deposits, fields and hedges; at another level it is something we inhabit, both physically and imaginatively. It is many-faceted, relying on an engagement with physical remains but also on emotional and aesthetic responses and on the power of memory, history and association.

Rodwell, Dennis, letter with formal response to English Heritage (3 August 2000).

Fairclough, Graham et al. (eds), *Europe's Cultural Landscape: Archaeologists and the Management of Change*, Europae Archaeologiae Consilium, Brussels (2002).

Historic Environment Review Steering Group, *Power of Place: The Future of the Historic Environment*, Power of Place Office, English Heritage, London (2000).

Preston, John, 'Power of Place: an agenda for the future?', in *News 16*, Institute of Historic Building Conservation (February 2001).

Local Authority World Heritage Forum, 'Power of Place: Response by LAWHF', LAWHF (March 2001).

Department for Culture, Media and Sport, *The Historic Environment: A Force for Our Future*, DCMS, London (2001).

English Heritage, *State of the Historic Environment Report 2002* (summary), English Heritage, London (2002).

English Heritage, *State of the Historic Environment Report 2002*, English Heritage, London (2002).

Rodwell, Dennis, letter with comments to English Heritage (26 February 2003).

English Heritage, *Conservation-led Regeneration: The Work of English Heritage*, English Heritage, London (1998).

English Heritage, *The Heritage Dividend: Measuring the Results of English Heritage Regeneration 1994–1999*, English Heritage, London (second edn, 2000); first published (1999).

English Heritage, *The Heritage Dividend 2002: Measuring the Results of English Heritage Regeneration 1999–2002*, English Heritage, London (2002).

Rodwell, Dennis, 'Industrial World Heritage sites in the United Kingdom', in *World Heritage Review*, UNESCO, Paris, and Ediciones San Marcos, Madrid (December 2002).

Television Education Network, 'Surveying Historic Buildings: What Every Surveyor Should Know' (programme notes:), TEN (February 2000). A sentence in these notes reads as follows: 'It is estimated that up to 75% of English Heritage grant aid is spent annually reversing unsuitable or damaging interventions carried out to historic buildings relatively recently.'

English Heritage, *Streamlining Listed Building Consent: Lessons from the use of Management Agreements: A Research Report*, English Heritage, London (2003).
Department for Culture, Media and Sport, *Protecting our Historic Environment: Making the System Work Better*, DCMS, London (2003).
Department for Culture, Media and Sport, *Review of Heritage Protection: The Way Forward*, DCMS, London (2004).
Department for Culture, Media and Sport, *Revisions to Principles of Selection for Listed Buildings: Planning Policy Guidance Note 15: Consultation Document*, Department for Culture, Media and Sport, London (2005).
English Heritage, *Heritage Counts 2005: The State of England's Historic Environment*, English Heritage, London (2005).
English Heritage, 'Heritage Protection Review', in *Conservation Bulletin*, English Heritage, London (Summer 2006).

Urban conservation: a cathedral city

Favier, Jean, *L'Univers de Chartres*, Bordas, Paris (1988).
Boucly, Yves (ed.), *Administration: Revue d'Information*, L'Association du corps préfectoral et des hauts fonctionnaires du Ministère de l'Intérieur, Paris (June 1971).
Rodwell, Dennis, 'Conservation on the Continent: A Report on the Protection of Historic Buildings and Urban Areas in France, Italy, Austria and Germany', unpublished, Cambridge (1972).
Rodwell, Dennis, 'The French connection: the significance for Edinburgh of conservation policies in the Marais, Paris', in: Harrison, Patrick (ed.), *Civilising the City: Quality or Chaos in Historic Towns*, Nic Allen, Edinburgh (1990).
Rodwell, Dennis, 'New light on the cities that sell their souls', in *Context*, Institute of Historic Building Conservation (June 2000).

Chapter 6

Sustainable Cities

Elkin, Timothy, et al., *Reviving the City: Towards Sustainable Urban Development*, Friends of the Earth, London (1991).
Breheny, Michael (ed.), *Sustainable Development and Urban Form*, Pion, London (1992).
Haughton, Graham and Hunter, Colin, *Sustainable Cities*, Routledge, London (reprinted, 2003); first published, Jessica Kingsley (1994).
Newman, Peter and Kenworthy, Jeffrey, *Sustainablity and Cities: Overcoming Automobile Dependence*, Island Press, Washington DC (1999).
Girardet, Herbert, *Creating Sustainable Cities*, Green Books, Totnes (1999).
Satterthwaite, David (ed.), *The Earthscan Reader in Sustainable Cities*, Earthscan Publications, London (1999).
Jenks, Mike et al. (ed.), *The Compact City: A Sustainable Urban Form?*, E & FN Spon, London (1996).
Williams, Katie et al. (ed.), *Achieving Sustainable Urban Form*, E & FN Spon, London (2000).
Rogers, Richard with Gumuchdjian, Philip, *Cities for a Small Planet*, Faber & Faber, London (1997).
Rogers, Richard and Powers, Anne, *Cities for a Small Country*, Faber & Faber, London (2000).
Brandon, Peter and Lombardi, Patrizia, *Evaluating Sustainable Development in the Built Environment*, Blackwell Publishing, Oxford (2005).

Rankine, Kerry, *Building the Future: A Guide to Building without PVC*, Greenpeace UK, London (1996).

Urban Villages

Charles, Prince of Wales, *A Vision of Britain: A Personal View of Architecture*, Doubleday, London (1989).
Rodwell, Dennis, 'Identity and Community' (book review), in *RSA Journal*, Royal Society of Arts, London (August/September 1992).
Aldous, Tony, *Urban Villages: A concept for creating mixed-use developments on a sustainable scale*, Urban Villages Group, London (1992).
Aldous, Tony (ed.), *Economics of Urban Villages*, Urban Villages Forum, London (1995).
Leccese, Michael and McCormick, Kathleen (eds), *Charter of the New Urbanism*, McGraw-Hill, New York (1999).
Congress for the New Urbanism, *Charter of the New Urbanism*, Congress for the New Urbanism, Chicago (2001).
Blake, William, quoted from the poem 'Jerusalem', written in 1804.
Thompson-Fawcett, Michelle, 'The Contribution of Urban Villages to Sustainable Development', in: Williams, Katie et al. (ed.), *Achieving Sustainable Urban Form*, E & FN Spon, London (2000).
Landman, Karina, 'Sustainable *Urban Village* Concept: Mandate, Matrix or Myth', paper presented at the conference *Technology and Management for Sustainable Building*, Pretoria, South Africa (May 2003).
Rosenau, Helen, *The Ideal City: Its Architectural Evolution in Europe*, Methuen, London (third edn, 1983); first published as: *The Ideal City in its Architectural Evolution*, Routledge and Kegan Paul, London (1959); second edn, Studio Vista, London (1974).
Lynch, Kevin, *Good City Form*, MIT Press, Cambridge, Massachusetts (1981).
Hall, Peter, *Cities of Tomorrow*, Blackwell, Oxford (updated edn, 1996); first published, Blackwell, Oxford (1988).
Biddulph, Mike, 'Villages Don't Make a City', in *Journal of Urban Design*, Routledge, London (vol 5, no 1, 2000).
Kidder Smith, George, *The New Architecture of Europe*, World Publishing, New York (1961).

Urban Renaissance

Rogers, Lord (chairman), *Towards an Urban Renaissance*, Urban Task Force, London (1999).
Rogers, Lord (chairman), *Towards an Urban Renaissance: Executive Summary*, Urban Task Force, London (1999).
Derby City Council, *Statutory List: Buildings of Special Architectural or Historic Interest*, Derby City Council, Derby (second edn, 1996).
Biddle, Gordon et al., *The Railway Heritage of Britain: 150 Years of Railway Architecture and Engineering*, Michael Joseph, London (1983).
Gehl, Jan, 'The Challenge of Making a Human Quality in the City', in *Perth Beyond 2000: A Challenge for a City*, Proceedings of the City Challenge Conference, Perth, Australia (1992).
Gehl, Jan and Gemzøe, Lars, *Public Spaces – Public Life: Copenhagen*, The Danish Architectural Press and The Royal Danish Academy of Fine Arts School of Architecture, Copenhagen (2004).
Newman, Peter, 'Local Symbolic Gestures to the Mainstream: Next Steps in Local Urban Sustainability', in *Local Environment*, Australia (vol 3, no 3, 1998).

Rogers, Lord (chairman), *Towards a Strong Urban Renaissance*, Urban Task Force, London (2005).
Gardiner, Joey, 'Urban areas being ignored, says task force', in *Regeneration & Renewal*, Haymarket Professional Publications, London (25 November 2005).
Taylor, Michael, 'Pathfinders at regeneration street', in *Context*, Institute of Historic Building Conservation (November 2003).
Owen-John, Harry, 'Pathfinders and the historic enviroment', in *Context*, Institute of Historic Building Conservation (November 2003).
Morgan, Ann et al., 'The Victorian Terrace: An Endangered Species Again?', in *The Victorian*, The Victorian Society, London (March 2006).
Wilkinson, Adam, *Pathfinder*, SAVE Britain's Heritage, London (2006).
Hall, Peter, 'Stricter targets will bring fewer homes', in *Regeneration & Renewal*, Haymarket Professional Publications, London (25 November 2005).

Empty properties in historic cities

Petherick, Ann et al., *Living Over The Shop: A Guide to the Provision of Housing Above Shops in Town Centres*, Joseph Rowntree Foundation, York (first report, June 1990).
Petherick, Ann et al., 'Housing Shortage? – Unlock Upstairs', in *Cornerstone*, SPAB, London (vol. 27, no. 2, 2006).
Department of the Environment, Transport and the Regions, *Our Towns and Cities: The Future: Delivering an Urban Renaissance* (the Urban White Paper), DETR, London (2000).

Urban conservation: a metropolitan centre

Boucly, Yves (ed.), *Administration: Revue d'Information*, L'Association du corps préfectoral et des hauts fonctionnaires du Ministère de l'Intérieur, Paris (June 1971).
Rodwell, Dennis, 'Conservation on the Continent: A Report on the Protection of Historic Buildings and Urban Areas in France, Italy, Austria and Germany', unpublished, Cambridge (1972).
Rodwell, Dennis, 'The French connection: the significance for Edinburgh of conservation policies in the Marais, Paris', in: Harrison, Patrick (ed.), *Civilising the City: Quality or Chaos in Historic Towns*, Nic Allen, Edinburgh (1990).
Kennet, Lord, 'Notes of a Visit to *Secteurs Sauvegardés* in France: June 1966', unpublished paper (1966).
Cantacuzino, Sherban (ed.), 'Paris', in *Architectural Review*, Architectural Press, London (September 1979).
Davy, Peter (ed.), 'Politics of Paris', in *Architectural Review*, Architectural Press, London (December 1986).
Rodwell, Dennis, 'Paris bouge: le Marais s'embourgeoise', unpublished paper (1998).
Pater, F, 'Shop Soiled', in *Architects' Journal*, Architectural Press, London (31 October 1990).

Chapter 7

International documentation

UNESCO, *Convention Concerning the Protection of the World Cultural and Natural Heritage* (the *World Heritage Convention*), UNESCO, Paris (1972).

UNESCO, *Operational Guidelines for the Implementation of the World Heritage Convention*, UNESCO, Paris (latest revision, February 2005); first published, UNESCO, Paris (1977).

UNESCO, *Recommendation Concerning the Protection, at National Level, of the Cultural and Natural Heritage*, UNESCO, Paris (1972).

Council of Europe, *European Charter of the Architectural Heritage* (the *European Charter*), Council of Europe (1975).

The Declaration of Amsterdam, adopted at the Congress on the European Architectural Heritage, Amsterdam (1975).

UNESCO, *Recommendation Concerning the Safeguarding and Contemporary Role of Historic Areas*, UNESCO, Nairobi (1976).

ICOMOS, *The Charter for the Conservation of Historic Towns and Urban Areas* (the *Washington Charter*), adopted at the ICOMOS General Assembly, Washington (1987).

Feilden, Bernard and Jokilehto, Jukka, *Management Guidelines for World Cultural Heritage Sites*, ICCROM, Rome (second edn, 1998); first published (1993).

Bath (in chronological order of publication)

Abercrombie, Patrick et al., *A Plan for Bath*, Isaac Pitman, Bath (1945).

Ison, Walter, *The Georgian Buildings of Bath from 1700 to 1830*, Faber and Faber, London (1948).

Coard, Peter and Ruth, *Vanishing Bath*, Kingsmead Press, Bath (1973).

Fergusson, Adam, *The Sack of Bath*, Compton Russell, Salisbury (1973).

Cantacuzino, Sherban (ed.), 'Bath: City in Extremis', in *Architectural Review*, Architectural Press, London (May 1973).

Department of Architecture and Planning, *Yesterday's Tomorrow: Conservation in Bath*, Bath City Council, Bath (1975).

Worskett, Roy, *Saving Bath: A Programme for Conservation*, Bath City Council, Bath (1978).

Pound, Christopher, *Genius of Bath: The City and its Landscape*, Millstream Books, Bath (1986).

ICOMOS, 'Advisory body evaluation: The City of Bath', ICOMOS, Paris (1987).

Fergusson, Adam and Mowl, Tim, *The Sack of Bath – And After*, Michael Russell, Salisbury (1989).

Wainwright, Martin, *The Bath Blitz*, DK Printing, Bath (third edn 1992); first published, Bath (1975).

Cunliffe, Barry, *The Roman Baths at Bath*, Bath Archaeological Trust, Bath (1993).

Bishop, Philippa et al., *Number 1 Royal Crescent Bath*, Bath Preservation Trust, Bath (1994).

Walcot, Biddy and James, Debbie, *The William Herschel Museum*, Bath Preservation Trust, Bath (undated, *c*.1995).

Department for Culture, Media and Sport, *World Heritage Sites: The Tentative List of the United Kingdom of Great Britain and Northern Ireland*, DCMS, London (1999).

Palladio, Andrea, *I Quattro Libri dell'Architettura*, Georg Olms, Hildesheim (1999); first edn 1570.

Woodward, Christopher, *The Building of Bath*, The Building of Bath Museum, Bath (undated, *c*.2000).

Planning Projects and Partnerships Team, *Living in a Conservation Area* (advice paper), Bath & North East Somerset Council, Bath (undated, *c*.2002).

Millington, John, *Beckford's Tower, Bath*, Bath Preservation Trust, Bath (seventh edn, 2002).

Bath & North East Somerset Council, *City of Bath World Heritage Site Management Plan: 2003–2009*, Bath & North East Somerset Council, Bath (2003). Apart from the Council itself, the ongoing steering group includes: Department for Culture, Media and Sport; English Heritage; ICOMOS-UK; The National Trust; Bath Federation of Residents Associations;

Bath Chamber of Commerce; Bath Preservation Society; Envolve (the sustainability charity); University of Bath; First Group; Bath Tourism Plus; and Avon and Somerset Constabulary. The stakeholder group comprises over a hundred local organisations and the general public.

Bath & North East Somerset Council, *City of Bath World Heritage Site Management Plan Summary: 2003–2009*, Bath & North East Somerset Council, Bath (2003).

Spence, Cathryn, *Obsession: John Wood and the Creation of Georgian Bath*, The Building of Bath Museum, Bath (2004).

McLaughlin, David, *Blitzed! War Artists in Bath*, Bath & North East Somerset Council, Bath (2005).

Rodwell, Dennis, 'City of Bath: A Masterpiece of Town Planning', in *World Heritage Review*, UNESCO, Paris, and Ediciones San Marcos, Madrid (October 2005).

Edinburgh (in chronological order of publication)

Lindsay, Ian, *Georgian Edinburgh*, Scottish Academic Press, Edinburgh (1973); this edition was revised posthumously by David Walker; first published, Edinburgh (1948).

Abercrombie, Patrick and Plumstead, Derek, *A Civic Survey and Plan for the City and Royal Burgh of Edinburgh*, Oliver and Boyd, Edinburgh (1949).

Youngson, Alexander, *The Making of Classical Edinburgh: 1750–1840*, Edinburgh University Press, Edinburgh (1966).

Young, Douglas et al., *Edinburgh in the Age of Reason*, Edinburgh University Press, Edinburgh (1967).

Begg, Ian and Denver, Boyd, *Edinburgh: The New Town*, Howie and Seath, Edinburgh (1967).

The Joint Consultants, 'City of Edinburgh Planning and Transport Study: Alternatives for Edinburgh: Summary' (leaflet), The Joint Consultants, Edinburgh (November 1971).

Scott, David, 'Electric rail system could save £60m, claims professor', in *The Scotsman*, Scotsman Publications, Edinburgh (20 March 1972).

Matthew, Sir Robert (ed.), *The Conservation of Georgian Edinburgh*, Edinburgh University Press, Edinburgh (1972).

Rodwell, Dennis, 'Urban Conservation in Edinburgh', unpublished, Cambridge (1972).

Rodwell, Dennis, 'Edinburgh lessons in conservation', in *The Daily Telegraph*, London (27 August 1973).

Bruce, George, *Some Practical Good: The Cockburn Association 1875–1975*, Cockburn Association, Edinburgh (1975).

McWilliam, Colin, *New Town Guide: The Story of Edinburgh's Georgian New Town*, Edinburgh New Town Conservation Committee, Edinburgh (1978).

Davey, Andy et al., *The Care and Conservation of Georgian Houses: A Maintenance Manual for the New Town of Edinburgh*, Paul Harris, Edinburgh (1978); since revised.

Rodwell, Dennis, 'Historic buildings face a crisis of maintenance', in *The Scotsman*, Scotsman Publications, Edinburgh (4 June 1980).

National Consumer Council, *What's Wrong with Walking*, HMSO, London (1987).

Harrison, Patrick (ed.), *Civilising the City: Quality or Chaos in Historic Towns*, Nic Allen, Edinburgh (1990).

Rodwell, Dennis, 'What about the children?', in *Context*, Association of Conservation Officers (September 1991).

Rodwell, Dennis, 'Civilising town centres', in *Context*, Association of Conservation Officers (September 1992).

City of Edinburgh District Council, *Central Edinburgh Local Plan: Draft Written Statement Approved for Consultation*, City of Edinburgh District Council, Edinburgh (1992).

Rodwell, Dennis, 'Edinburgh – A European City of the Twenty First Century: New Thinking for a New Millennium – A Sustainable Vision', local plan consultation paper, Edinburgh (1992).

ICOMOS, 'Advisory body evaluation: The Old and New Towns of Edinburgh', ICOMOS, Paris (1995).

Editorial, 'Transport of limited delights', in *The Scotsman*, Scotsman Publications, Edinburgh (16 August 1996).

Riddock, Lesley, 'Fifty years of muddle', in *The Scotsman*, Scotsman Publications, Edinburgh (16 August 1996).

Johnson, Jim, 'Sustainability in Scottish Cities', in: Jenks, Mike et al. (ed.), *The Compact City: A Sustainable Urban Form?*, E & FN Spon, London (1996).

Rodwell, Dennis, 'Balance and involvement in a world heritage city', in *Context*, Institute of Historic Building Conservation (December 2000).

Hendry, Arnold, *The Edinburgh Transport Saga*, Whitehouse Publications, Edinburgh (2002).

Harrison, Patrick, *Urban Pride: Living and Working in a World Heritage City*, Edinburgh World Heritage Trust, Edinburgh (2002).

Rodwell, Dennis, 'From accolade to responsibility', in *Context*, Institute of Historic Building Conservation (November 2002).

Rodwell, Dennis, 'Exemplary practice in the rehabilitation of historic cities', paper delivered at the UNESCO international seminar *Management of Private Property in the Historic City-Centres of the European Countries-in-Transition*, Pécs, Hungary (November 2002).

Campbell, Donald, *Edinburgh: A Cultural and Literary History*, Signal Books, Oxford (2003).

Dalton, Alastair, 'Charging ahead to tackle gridlock', in *The Scotsman*, Scotsman Publications, Edinburgh (12 September 2003).

Edinburgh Tourism Action Group, *Edinburgh Tourism Action Plan: 04/07*, Edinburgh Tourism Action Group, Edinburgh (2004).

Edinburgh World Heritage, *Edinburgh World Heritage Site Management Plan: Public Consultation Draft – September 2004*, Edinburgh World Heritage, Edinburgh (2004).

Rodwell, Dennis, 'Edinburgh: Responsibilities of a World Heritage Site and of a World Heritage City', paper delivered to the Cockburn Association, Edinburgh (2004).

Rodwell, Dennis, 'City of Edinburgh World Heritage Site Management Plan Public Consultation Draft – September 2004: Comments', consultation response, Edinburgh (2004).

Wainscoat, Nancy, 'Edinburgh rejects city entry charge', in *Planning*, Haymarket Professional Publications, London (25 February 2005).

Wainscoat, Nancy, 'Road charge rejection triggers policy rethink', in *Planning*, Haymarket Professional Publications, London (4 March 2005).

Edinburgh City Centre Management Company, *Inspiring Action – The Edinburgh City Centre Action Plan: 2005–10*, Edinburgh City Centre Management Company, Edinburgh (2005).

City of Edinburgh Council, *Edinburgh City Centre: Charting the Way Forward*, City of Edinburgh Council, Edinburgh (final version, May 2005).

Edinburgh World Heritage, *The Old and New Towns of Edinburgh World Heritage Site Management Plan*, Edinburgh World Heritage, Edinburgh (July 2005). The steering group comprised: City of Edinburgh Council, Scottish Executive Planning Divisions, and Historic Scotland (an executive agency within the Scottish Executive). The main parties listed as responsible for managing the site are: Edinburgh World Heritage; Historic Scotland; the City of Edinburgh Council; Scottish Enterprise Edinburgh and Lothian; and the Edinburgh City Centre Management Company.

Glendinning, Miles, 'The *Grand Plan*: Robert Matthew and the Triumph of Conservation in Scotland, in *Architectural Heritage XVI*, Architectural Heritage Society of Scotland, Edinburgh (2005).

Edwards, Brian and Jenkins, Paul (eds), *Edinburgh: The Making of a Capital City*, Edinburgh University Press, Edinburgh (2005).

McCall Smith, Alexander et al., *One City*, Polygon, Edinburgh (2005).

Robinson, David, 'The bullet-proof barrier between city's rich and poor', in *The Scotsman*, Scotsman Publications, Edinburgh (3 December 2005).

Edinburgh World Heritage, *Edinburgh World Heritage Business Plan (2006–07)*, Edinburgh World Heritage, Edinburgh (January 2006).

Edinburgh World Heritage, *The Old and New Towns of Edinburgh World Heritage Site Draft Action Programme*, Edinburgh World Heritage, Edinburgh (March 2006).

Sabadus, Aura, 'Edinburgh at risk of becoming *clone city*, says Charles', in *The Scotsman*, Scotsman Publications, Edinburgh (1 June 2006).

Bath and Edinburgh (in chronological order of publication)

Rodwell, Dennis, 'The World Heritage Convention and the Exemplary Management of Complex Heritage Sites', in *Journal of Architectural Conservation*, Donhead Publishing, Donhead St Mary (November 2002).

Rodwell, Dennis, 'Sustainability and the Holistic Approach to the Conservation of Historic Cities', in *Journal of Architectural Conservation*, Donhead Publishing, Donhead St Mary (March 2003).

Rodwell, Dennis, 'Managing Historic Cities: the Management Plans for the Bath and Edinburgh World Heritage Sites', in *Journal of Architectural Conservation*, Donhead Publishing, Donhead St Mary (July 2006).

Rodwell, Dennis, 'The values of world heritage', in *Context*, Institute of Historic Building Conservation (July 2006).

Chapter 8

Central and Eastern Europe

Rodwell, Dennis, *The Revitalisation of World Heritage Cities in Central and Eastern Europe* (mission report), World Heritage Centre, UNESCO, Paris (May 1999).

Samol, Frank, 'Urban Rehabilitation in Central and Eastern Europe', paper delivered at the *International Symposium on the Rehabilitation of Historic Cities in Eastern and South Eastern Europe*, Sibiu (October 2002).

Rodwell, Dennis, 'Approaches to Urban Conservation in Central and Eastern Europe', in *Journal of Architectural Conservation*, Donhead Publishing, Donhead St Mary (July 2003).

Czyzewska, Aleksandra, 'The condominium association loan programme in Sopot, Poland', in: UNESCO, *Management of Private Property in the Historic City-Centres of the European Countries-in-Transition* (proceedings of UNESCO international seminar, Bucharest, April 2001), Division of Cultural Heritage, UNESCO, Paris (2002).

Skalski, Krzysztof, *O budowie systemu rewitalizacji dawnych dzielnic miejskich* (*Constitution d'un système de réhabilitation d'anciens quartiers urbains; aspects structurels sur la base de l'expérience française*), Krakowski Instytut Nieruchomości, Krakov (1996).

Sibiu (in chronological order of publication)

Giurescu, Dinu, *The Razing of Romania's Past*, ICOMOS-US, Washington DC (1990).

Fabini, Hermann and Alida, *Hermannstadt: Porträt einer Stadt in Siebenbürgen*, Monumenta Verlag, Heidelberg (2000).

German Agency for Technical Cooperation (on behalf of the City Hall, Sibiu), *Charter for the Rehabilitation of the Historic Center of Sibiu/Hermannstadt*, GTZ, Sibiu (second edn, October 2000).

Groep Planning, *Sibiu/Hermannstadt, Romania: Towards a Sustainable Rehabilitation and Development Plan*, Groep Planning, Bruges (2001).

Avram, Alexandru, *Sibiu: History and Monuments*, Global Media, Sibiu (2001).

German Agency for Technical Cooperation (on behalf of the City Hall, Sibiu), *Kommunales Aktionsprogramm: Sibiu 2001–2004*, GTZ, Sibiu (2001).

Nistor, Sergiu, 'Romania's Heritage: between Neglect and Revitalization: Bridging the Gap towards Sustainable Management of Historic Properties: The Role of International Co-operation', in: UNESCO, *Management of Private Property in the Historic City-Centres of the European Countries-in-Transition* (proceedings of UNESCO international seminar, Bucharest, April 2001), Division of Cultural Heritage, UNESCO, Paris (2002).

Romanian-German Cooperation Project, *International Symposium on the Rehabilitation of Historic Cities in Eastern and South Eastern Europe* (symposium report), Sibiu City Hall and GTZ, Sibiu (2002).

Rodwell, Dennis, 'From accolade to responsibility', in *Context*, Institute of Historic Building Conservation (November 2002).

Rodwell, Dennis, 'Exemplary practice in the rehabilitation of historic cities', paper delivered at the UNESCO international seminar *Management of Private Property in the Historic City-Centres of the European Countries-in-Transition*, Pécs, Hungary (November 2002).

Rodwell, Dennis, 'Management of Historic Cities in the European Countries-in-Transition: experience and problems', paper delivered at the conference *Continuity of Urban Development in Historic Cities*, Vilnius, Lithuania (June 2003).

Rodwell, Dennis and Nistor, Sergiu, 'World Heritage Convention: Tentative List for Romania: The Historic Centre of Sibiu' (final draft), Sibiu (3 December 2003).

German Agency for Technical Cooperation (on behalf of the City Hall, Sibiu), *Sibiu Historic Centre Management Plan: 2005–2009*, GTZ, Sibiu (2005).

Niedermaier, Paul (ed.), *Nomination of the Sibiu Historic Centre for Inscription on the World Heritage List*, Institutul de Cercetări Socio-Umane Sibiu, Sibiu (September 2005).

Community involvement elsewhere in Romania

Baxter, David, 'Prince of Wales inspects IHBC work in Transylvania', in *Context*, Institute of Historic Building Conservation (July 2002).

Baxter, David, 'Conservation training at Banffy Castle makes its mark', in *Context*, Institute of Historic Building Conservation (November 2003).

Baxter, David, 'New directions for training at Banffy Castle', in *Context*, Institute of Historic Building Conservation (March 2005).

Baxter, David, 'Conservation training at Banffy Castle', in *Context*, Institute of Historic Building Conservation (November 2006).

Rodwell, Dennis, 'Community involvement in the regeneration of a historic city centre', in: Carman, John and White, Roger (eds), *World Heritage: Global Challenges and Local Solutions*, British Archaeological Reports, International Series, Archaeopress, Oxford (forthcoming).

Asmara (in chronological order of publication)

Siravo, Francesco, *Preservation and Presentation of the Cultural Heritage – Asmara and Massawa*, UNESCO, Paris (1995).

Municipality of Asmara, *Building Regulations*, Municipality of Asmara, Asmara (1998); this is the translation into English by Rita Mazzocchi-Dawkins of: *Regolamento Edilizio*, Asmara (1938).

Oriolo, Leonardo (ed.), *Asmara Style*, Scuola Italiana, Asmara (1998).

Denison, Edward and Paice, Edward, *The Bradt Travel Guide to Eritrea*, Bradt Travel Guides, Chalfont St Peter, England (third edn, 2002).

Hill, Justin, *Ciao Asmara*, Abacus, London, 2002.

Denison, Edward and Ren, Guang Yu, *Asmara Architecture Archives: Final Report*, Cultural Assets Rehabilitation Project, Asmara (August 2002).

Denison, Edward et al., *Asmara: Africa's Secret Modernist City*, Merrell, London and New York (2003).

Gebremedhin, Naigzy et al, *Asmara: A Guide to the Built Environment*, Cultural Assets Rehabilitation Project, Asmara (2003).

Municipality of Asmara and Cultural Assets Rehabilitation Project, *Asmara City Map & Historic Perimeter*, Asmara (2003).

Caesar, Karin and Rosengren, Katarina, *An Analysis of the Situation for Cyclists in Asmara with Emphasis on Safety Aspects* (master's thesis), Department of Technology and Society, Lund University, Sweden (January 2003).

Lund team, *Building Permits and Planning Regulations in Asmara: A Twinning Project*, Lund, Sweden (February 2003).

Lund team, *Bicycle Traffic Structure Plan for Asmara: Detailed drawings for a Golden Lane*, Lund, Sweden (February 2003).

Ghebrikadan, Alemseghed, *Organising Heritage Tourism: Scheme for Safeguarding of Urban Heritage and Attractions*, Asmara (February 2004).

Brivio, Peppio et al., *Planning Initiative for the Historic Perimeter of Asmara*, Cultural Assets Rehabilitation Project, Asmara (June 2003).

Wischmeijer, J. and T., *Report on Missions 28427 and 28429M ER to the Cultural Assets Rehabilitation Project, 17 July to 7 August 2003*, Cultural Assets Rehabilitation Project, Asmara (August 2003).

Ghebray, Amanuel, *Survey and Documentation of the Historic Perimeter of Asmara* (prepared for the Department of Infrastructure Services, Administration of Zoba Maakel), Eritrean Consulting for the Horn of Africa, Asmara (August 2003).

Lacey, Marc, 'In Asmara all roads lead to Rome', in *International Herald Tribune*, New York Times, Paris (16 September 2003).

Denison, Edward and Ren, Guang Yu, 'Supporting Africa's Secret City', in *The Globalist*, (7 February 2004).

Robinson, Simon, 'Africa's Art Deco Capital', in *Time* (16 February 2004).

Street, Mike, 'Asmara: From Dreams to Reality: Part 1', in *Eritrea Profile*, Asmara (20 March 2004).

Van Grinsven, Jan, *Report on Mission 30233 M ER to the Cultural Assets Rehabilitation Project, 17 February to 2 March 2004*, Cultural Assets Rehabilitation Project, Asmara (March 2004).

Rodwell, Dennis, *Over-arching Urban Planning and Building Conservation Guidelines for the Historic Perimeter of Asmara, Eritrea* (mission report), Cultural Assets Rehabilitation Project, Asmara (March 2004).

Rodwell, Dennis, 'Asmara: Conservation and Development in a Historic City', in *Journal of Architectural Conservation*, Donhead Publishing, Donhead St Mary (November 2004).

Chapter 9

Think global, act local

Mehaffy, Michael (ed.), *The Order of Nature: The City as Ecosystem* (speaker position papers), The Prince's Foundation, London (September 2005).
Brand, Janet (ed.), *Planning for Sustainable Development: Conference Papers*, The Planning Exchange (1993).
Buchanan, Colin, *Traffic in Towns*, HMSO, London (1963).
Cleere, Henry, *ICOMOS-UK International Briefing*, ICOMOS-UK (Autumn 2006).
Bortolotto, Chiara, 'From cultural objects to cultural processes: UNESCO's intangible cultural heritage', in *Journal of Museum Ethnography*, Museum Ethnographers Group (forthcoming).
Rodwell, Dennis, 'The determining role of Historic Cities in the quest for Sustainable Urban Development', paper delivered at the *Heritage Forum*, London (June 2000).
Cullen, Gordon, *Townscape*, Architectural Press, London (1961).

Reduce, reuse and recycle

Strong, David, 'Sustainable Buildings – Time for a Radical Rethink?', paper delivered at the conference *The Order of Nature: The City as Ecosystem*, The Prince's Foundation, London (September 2005).
Booth, Robert, 'Green Shoots of Sustainability', in *Building Design*, CMP Information, London (15 October 2004).
Birch, Amanda, *Green Paper: Architecture and Sustainability*, in *Building Design*, CMP Information, London (October 2004); published as a supplement to the edition of 15 October 2004.
Jokilehto, Jukka, *A History of Architectural Conservation*, Elsevier Butterworth-Heinemann, Oxford (2004); first published, Butterworth-Heinemann, Oxford (1999).
Rodwell, Dennis, 'A house in the country', in *Scottish Field*, Holmes McDougall, Glasgow (June 1985).
Greysmith, Brenda, 'Little house on the Green', in *Traditional Homes*, Benn, St Albans (April 1987).
Cantacuzino, Sherban, *New Uses for Old Buildings*, Architectural Press, London (1975).
SAVE Britain's Heritage, *Catalytic Conversion: Revive Historic Buildings to Regenerate Communities*, SAVE Britain's Heritage, London (1998).
Latham, Derek, *Creative Re-Use of Buildings*, Donhead Publishing, Donhead St Mary (vols 1 and 2, 2000).
Rodwell, Dennis, 'What does one do with a dilapidated pigeon coop?', in *Scottish Field*, Holmes McDougall, Glasgow (August 1986).
Rodwell, Dennis, 'A terminal state: the rescue of Melrose Station House', in *Context*, Association of Conservation Officers (January 1989).
Caplan, Neil, *The Waverley Route*, Ian Allan, Weybridge (1985).
Rodwell, Dennis, *The Bolshoi Theatre, Moscow, Russia* (mission report), Division of Cultural Heritage, UNESCO, Paris (2003).
Charlton, Christopher (ed.), *The Derwent Valley Mills and their Communities*, Derwent Valley Mills Partnership, Matlock, (2001).

ICOMOS, 'Advisory body evaluation: Derwent Valley Mills, Derbyshire, United Kingdom', ICOMOS, Paris (2001).
Rodwell, Dennis, *Darley Abbey Conservation Area Appraisal*, Derby City Council, Derby (revised for publication, 2003; publication forthcoming).
Rodwell, Dennis, *Darley Abbey Mills: Condition Survey – Initial Overview*, Derby City Council, Derby (2000).
Tetlow, John and Goss, Anthony, *Home, Towns and Traffic*, Faber and Faber, London (second edn, 1968); first published, Faber and Faber, London (1965).
Rodwell, Dennis, 'Stone cleaning in urban conservation', in: Webster, Robin (ed.) *Stone Cleaning*, Donhead Publishing, London (1992).
Rodwell, Dennis, 'Anti-stonecleaning myopia', in *Context*, Association of Conservation Officers (December 1993).
Rodwell, Dennis, 'The periodic removal of dirt', in *Stone Industries*, Herald House, Worthing (December 1993/January 1994).

Top-down meeting bottom-up

Jacobs, Jane, *The Death and Life of Great American Cities: The Failure of Town Planning*, Random House, New York (1961).
Rodwell, Dennis, 'Sustainability and the Holistic Approach to the Conservation of Historic Cities', in *Journal of Architectural Conservation*, Donhead Publishing, Donhead St Mary (March 2003).

Urban conservation: expanding threats to historic cities

Shankland, Graeme, 'Conservation through Planning', in: Ward, Pamela (ed.), *Conservation and Development in Historic Towns and Cities*, Oriel Press, Newcastle-upon-Tyne (1968).
Rodwell, Dennis, *The Revitalisation of World Heritage Cities in Central and Eastern Europe* (mission report), World Heritage Centre, UNESCO, Paris (May 1999).
Rodwell, Dennis, 'Approaches to Urban Conservation in Central and Eastern Europe', in *Journal of Architectural Conservation*, Donhead Publishing, Donhead St Mary (July 2003).
ICOMOS, 'Advisory body evaluation: Historic Town of Banská Štiavnica and the Technical Monuments in its Vicinity, Slovak Republic', ICOMOS, Paris (1993).
ICOMOS, 'Advisory body evaluation: Historic Centre of Prague, Czech Republic', ICOMOS, Paris (1992).
ICOMOS, 'Advisory body evaluation: Historic Centre of Sighişoara, Romania', ICOMOS, Paris (1999).
Fejérdy, Tamás (Chairperson of the World Heritage Committee), letter to Ion Iliescu, President of Romania (5 February 2003).
Rae, Isobel, *Charles Cameron: Architect to the Court of Russia*, Elek Books, London (1971).
ICOMOS, 'Advisory body evaluation: Historic Centre of Leningrad and Surroundings, USSR', ICOMOS, Paris (1990).
Meetings with Sergei Ivanovich Sokolov, Chief Architect to the city of St Petersburg, in 1991 and 2004. In the majority of the countries of Central and Eastern Europe the title architect encompasses the qualification of urban planner.
Skibinsky, Sveta, 'Historic Center may be Shrunk', in *The St Petersburg Times*, St Petersburg, Russia (15 March 2005).
Haughton, Graham and Hunter, Colin, *Sustainable Cities*, Routledge, London (reprinted, 2003); first published, Jessica Kingsley (1994).
Denison, Edward, Ren, Guang Yu, *Building Shanghai: The Story of China's Gateway*, Wiley-Academy, Chichester (2006).

Chapter 10

Loyer, François, *Ville d'hier, ville aujourd'hui en Europe*, Editions du Patrimoine, Fayard, Paris (2001).
Rodwell, Dennis, 'New light on the cities that sell their souls', in *Context*, Institute of Historic Building Conservation (June 2000).
Cantacuzino, Sherban (ed.), *Architectural Conservation in Europe*, Architectural Press, London (1975).
Jokilehto, Jukka, *A History of Architectural Conservation*, Elsevier Butterworth-Heinemann, Oxford (2004); first published, Butterworth-Heinemann, Oxford (1999).
Nordic World Heritage Office, *Sustainable Historic Cities: A North-Eastern European Approach*, Nordic World Heritage Office, Oslo (Final Report, December 1998).
Loyer, François and Schmuckle-Mollard, Christiane, *Façadisme et Identité Urbaine*, Editions du Patrimoine, Paris (2001).
Feilden, Bernard, *Conservation of Historic Buildings*, Architectural Press, London (third edn, 2003); first published, Architectural Press, London (1982).
Richards, James (ed.), *Architectural Review*, Architectural Press, London (December 1970).
Pater, F, 'Shop Soiled', in *Architects' Journal*, Architectural Press, London (31 October 1990).
Urban and Economic Development Group, *Vital and Viable Town Centres: Meeting the Challenge*, HMSO, London (1994).
Rodwell, Dennis, 'Convenience of Location' (book review), in *RSA Journal*, Royal Society of Arts, London, (January/February 1995).
Kochan, Ben and Cane, Mary, 'Fitting New Shops into Traditional City Centres', in *Urban Environment Today*, Landor, London (24 January 2002).
Taylor, Craig, 'Where am I? – Why are our towns all the same?', in *The Guardian Weekend*, Guardian Newspapers, London (23 November 2002).
Willis, Ben, 'Market towns *losing character*', in *Regeneration & Renewal*, Haymarket Professional Publications, London (27 June 2003).
UNESCO, *Vienna Memorandum on World Heritage and Contemporary Architecture – Managing the Historic Urban Landscape*, World Heritage Centre, UNESCO, Paris (2005).
Gates, Charles, 'UN plots icon ban', in *Building Design*, CMP Information, London (13 May 2005).
Rodwell, Dennis, 'Urban conservation in an age of globalisation', in *Context*, Institute of Historic Building Conservation (November 2006).
Fleig, Karl, *Alvar Aalto*, Edition Girsberger, Zurich (1963).
Rodwell, Dennis, 'Conservation on the Continent: A Report on the Protection of Historic Buildings and Urban Areas in France, Italy, Austria and Germany', unpublished, Cambridge (1972).
Jodido, Philip, *Grands Travaux*, Connaissance des Arts, Paris (1992).
Bennet, Ellen, 'Prescott: What urban task force?', in *Building Design*, CMP Information, London (9 December 2005).
English Heritage and the Commission for Architecture and the Built Environment, *Building in Context: New Development in Historic Areas*, English Heritage/CABE, London (2001).
Rodwell, Dennis, 'The achievement of exemplary practice in the protection of our built heritage: the need for a holistic conservation- and sustainability-orientated vision and framework', *Management of Private Property in the Historic City-Centres of the European Countries-in-Transition*, Division of Cultural Heritage, UNESCO, Paris (2002).
Kulich, Jan, *Kutna Horá: St Barbara's Church and the Town*, Vega-L, Jižní (1998).
Fleming, John, *Robert Adam and his Circle*, John Murray, London (1962).
Beard, Geoffrey, *The Work of Robert Adam*, John Bartholomew, Edinburgh (1978).

Bryant, Julius, *Robert Adam: Architect of Genius*, English Heritage, London (1992).

Smith, John, 'How Much Should We Respect the Past?', in *Architectural Heritage III*, Edinburgh University Press, Edinburgh (1992).

Slavid, Ruth, 'Maintaining a sense of history', in *Architects Journal*, Architectural Press, London (5 July 2001).

Adam, Robert, 'Does heritage dogma destroy living history?', in *Context*, Institute of Historic Building Conservation (May 2003).

Rodwell, Dennis, 'Philosophy and practice' (published as a letter), in *Context*, Institute of Historic Building Conservation (July 2003).

ICOMOS-Hungary, *The Pécs Declaration on the Venice Charter*, ICOMOS-Hungary, Pécs (2004).

Dunlop, Alan, 'Lost Opportunity', in *Prospect*, Royal Incorporation of Architects in Scotland, Edinburgh (Summer 2006).

Rodwell, Dennis, 'Preserving and enhancing', in *Context*, Association of Conservation Officers (December 1992).

Rodwell, Dennis, 'Sustainability and the Holistic Approach to the Conservation of Historic Cities', in *Journal of Architectural Conservation*, Donhead Publishing, Donhead St Mary (March 2003).

Bibliography

The following list of principal documents is extracted from the list of sources attributable to each chapter, with selected additions.

General reference sources

Magnusson, Magnus (ed.), *Chambers Biographical Dictionary*, Chambers, Edinburgh (fifth edn, re-printed 1990).
Wikipedia: the free internet encyclopaedia. The main page may be found on: http://en.wikipedia.org/wiki/Main_Page
UNESCO World Heritage Centre: http://whc.unesco.org/

Charters and related documents (in chronological order)

Morris, William, *Manifesto of the Society for the Protection of Ancient Buildings* (the *SPAB Manifesto*), SPAB, London (1877).
The Athens Charter for the Restoration of Historic Monuments (the *Athens Charter*), Adopted at the First International Congress of Architects and Technicians of Historic Monuments, Athens (1931).
La Charte d'Athènes, Adopted at the Fourth CIAM (Congrès Internationaux d'Architecture Moderne) Congress, Paris (1933).
International Charter for the Conservation and Restoration of Monuments and Sites (the *Venice Charter*), Adopted at the Second International Congress of Architects and Technicians of Historic Monuments, Venice (1964).
Council of Europe, *European Charter of the Architectural Heritage* (the *European Charter*), Council of Europe (1975).
The Declaration of Amsterdam, Adopted at the Congress on the European Architectural Heritage, Amsterdam (1975).
ICOMOS-German Democratic Republic, *The Declaration of Dresden*, ICOMOS-GDR, Dresden (1982).
ICOMOS, *The Charter for the Conservation of Historic Towns and Urban Areas* (the *Washington Charter*), Adopted at the ICOMOS General Assembly, Washington (1987).
Lemaire, Raymond and Stovel, Herb (eds), *Nara Document on Authenticity*, Nara, Japan (1994).

ICOMOS-Australia, *The Australia ICOMOS Charter for Places of Cultural Significance* (the *Burra Charter*), ICOMOS-Australia (revision, November 1999); first published, ICOMOS-Australia (1979).
ICOMOS-Hungary, *The Pécs Declaration on the Venice Charter*, ICOMOS-Hungary, Pécs (2004).

UNESCO and related publications, including conventions and recommendations (in chronological order)

UNESCO, *Convention Concerning the Protection of the World Cultural and Natural Heritage* (the *World Heritage Convention*), UNESCO, Paris (1972).
UNESCO, *Recommendation Concerning the Protection, at National Level, of the Cultural and Natural Heritage*, UNESCO, Paris (1972).
UNESCO, *Recommendation Concerning the Safeguarding and Contemporary Role of Historic Areas*, UNESCO, Nairobi (1976).
Pressouyre, Léon, *The World Heritage Convention, Twenty Years Later*, UNESCO, Paris (1996).
Feilden, Bernard and Jokilehto, Jukka, *Management Guidelines for World Cultural Heritage Sites*, ICCROM, Rome (second edn, 1998); first published (1993).
UNESCO, 'Mission Statement' and associated set of papers, the World Heritage Centre, UNESCO, Paris (May 2000).
UNESCO, *Convention for the Safeguarding of the Intangible Cultural Heritage*, UNESCO, Paris (2003).
Batisse, Michel and Bolla, Gerard, *The Invention of 'World Heritage'*, Association of Former UNESCO Staff Members, Paris (2005).
UNESCO, *Operational Guidelines for the Implementation of the World Heritage Convention*, UNESCO, Paris (latest revision, February 2005); first published, UNESCO, Paris (1977).
UNESCO, *Vienna Memorandum on World Heritage and Contemporary Architecture – Managing the Historic Urban Landscape*, World Heritage Centre, UNESCO, Paris (2005).

Miscellaneous publications (in alphabetical order by author)

Aldous, Tony, *Urban Villages: A Concept for Creating Mixed-use Developments on a Sustainable Scale*, Urban Villages Group, London (1992).
Aldous, Tony (ed.), *Economics of Urban Villages*, Urban Villages Forum, London (1995).
Arnold, John et al. (eds), *History and Heritage: Consuming the Past in Contemporary Culture*, Donhead Publishing, Donhead St Mary (1998).
Arondel, Mathilde, *Chronologie de la politique urbaine: 1945–2000*, ANAH, Paris (2001).
Ashurst, John and Ashurst, Nicola, *Practical Building Conservation: English Heritage Technical Handbook*, Gower Technical Press, London (Volumes 1 to 5, 1988).
Bardauskienė, Dalia et al. (eds), *Vilnius Old Town Revitalisation: 1998–2003*, Jsc Ikstrys, Vilnius (2003).
Bath & North East Somerset Council, *City of Bath World Heritage Site Management Plan: 2003–2009*, Bath & North East Somerset Council, Bath (2003).
Bell, Colin and Rose, *City Fathers: The Early History of Town Planning in Britain*, Penguin Books, Harmondsworth (1972); first published by Barrie and Rockliff, Cresset Press (1969).
Benevolo, Leonardo, *The Origins of Modern Town Planning*, Routledge and Kegan Paul, London (1957); first published as: *Le Origini dell'Urbanistica Moderna*, Laterza, Bari (1963).

Binney, Marcus, *Our Vanishing Heritage*, Arlington Books, London (1984).
Boardman, Philip, *The Worlds of Patrick Geddes: Biologist, Town Planner, Re-educator, Peace-Warrior*, Routledge and Kegan Paul, London (1978).
Bortolotto, Chiara, 'From cultural objects to cultural processes: UNESCO's intangible cultural heritage', in *Journal of Museum Ethnography*, Museum Ethnographers Group (forthcoming).
Brand, Janet (ed.), *Planning for Sustainable Development: Conference Papers*, The Planning Exchange (1993).
Brandon, Peter and Lombardi, Patrizia, *Evaluating Sustainable Development in the Built Environment*, Blackwell Publishing, Oxford (2005).
Breheny, Michael (ed.), *Sustainable Development and Urban Form*, Pion, London (1992).
Briggs, Asa, *Victorian Cities*, Odhams, London (1963).
Bruce, George, *Some Practical Good: The Cockburn Association 1875–1975*, Cockburn Association, Edinburgh (1975).
Buchanan, Colin, *Traffic in Towns*, HMSO, London (1963).
Buchanan, Colin, *Traffic in Towns: The Specially Shortened Edition of the Buchanan Report*, Penguin Books, Harmondsworth (1964).
Buchanan, Colin and Partners, *Bath: A Study in Conservation*, HMSO, London (1968).
Burns, Wilfred, *New Towns for Old: The Technique of Urban Renewal*, HMSO, London (1963).
Burrows, G., *Chichester: A Study in Conservation*, HMSO, London (1968).
Burtenshaw, David et al., *The European City: A Western Perspective*, David Fulton, London (1991).
Cantacuzino, Sherban (ed.), *Architectural Conservation in Europe*, Architectural Press, London (1975).
Cantacuzino, Sherban, *New Uses for Old Buildings*, Architectural Press, London (1975).
Carman, John and White, Roger (eds), *World Heritage: Global Challenges and Local Solutions*, British Archaeological Reports, International Series, Archaeopress, Oxford (forthcoming).
Charles, Prince of Wales, *A Vision of Britain: A Personal View of Architecture*, Doubleday, London (1989).
Charlton, Christopher (ed.), *The Derwent Valley Mills and their Communities*, Derwent Valley Mills Partnership, Matlock, (2001).
Cherry, Gordon, *Cities and Plans: The Shaping of Britain in the Nineteenth and Twentieth Centuries*, Edward Arnold, London (1988).
Chieng, Diana Chan, *Projets Urbains en France*, Le Moniteur, Paris (2002).
Choay, Françoise, *L'urbanisme, Utopies et Réalités: Une Anthologie*, Seuil, Paris (1965).
Choay, Françoise (ed.), *La Conférence d'Athènes sur la conservation artistique et historique des monuments*, Editions de l'Imprimeur, Paris (2002).
Coleman, Alice, *Utopia on Trial: Vision and Reality in Planned Housing*, Hilary Shipman, London (1985).
Cormack, Patrick, *Heritage in Danger*, New English Library, London (1976).
Cullen, Gordon, *Townscape*, Architectural Press, London (1961).
Cullingworth, Barry and Nadin, Vincent, *Town & Country Planning in the UK*, Routledge, London (twelfth edn, 1997); first published (1964).
Davey, Andy et al., *The Care and Conservation of Georgian Houses: A Maintenance Manual for the New Town of Edinburgh*, Paul Harris, Edinburgh (1978); since revised.
Denison, Edward et al., *Asmara: Africa's Secret Modernist City*, Merrell, London and New York (2003).
Denison, Edward, Ren, Guang Yu, *Building Shanghai: The Story of China's Gateway*, Wiley-Academy, Chichester (2006).
Department for Culture, Media and Sport, *World Heritage Sites: The Tentative List of the United Kingdom of Great Britain and Northern Ireland*, DCMS, London (1999).

Department for Culture, Media and Sport, *The Historic Environment: A Force for Our Future*, DCMS, London (2001).
Department for Culture, Media and Sport, *Protecting our Historic Environment: Making the System Work Better*, DCMS, London (2003).
Department for Culture, Media and Sport, *Review of Heritage Protection: The Way Forward*, DCMS, London (2004).
Department of the Environment, Transport and the Regions, *Our Towns and Cities: The Future: Delivering an Urban Renaissance* (the Urban White Paper), DETR, London (2000).
Dresner, Simon, *The Principles of Sustainability*, Earthscan Publications, London (2002).
Earl, John, *Building Conservation Philosophy*, Donhead Publishing, Donhead St Mary (third edn, 2003); first published, College of Estate Management, Reading (1996); second edn (1997).
Edinburgh World Heritage, *The Old and New Towns of Edinburgh World Heritage Site Management Plan*, Edinburgh World Heritage, Edinburgh (July 2005).
Edinburgh World Heritage, *The Old and New Towns of Edinburgh World Heritage Site Draft Action Programme*, Edinburgh (March 2006).
Elkin, Timothy, et al., *Reviving the City: Towards Sustainable Urban Development*, Friends of the Earth, London (1991).
Elliott, Jennifer, *An Introduction to Sustainable Development*, Routledge, London (second edn, 1999); first published, Routledge, London (1994).
English Heritage, *Conservation-led Regeneration: The Work of English Heritage*, English Heritage, London (1998).
English Heritage and the Commission for Architecture and the Built Environment, *Building in Context: New Development in Historic Areas*, English Heritage/CABE, London (2001).
English Heritage, *State of the Historic Environment Report 2002*, English Heritage, London (2002).
English Heritage, *The Heritage Dividend 2002: Measuring the Results of English Heritage Regeneration 1999–2002*, English Heritage, London (2002).
English Heritage, *Guidance on the Management of Conservation Areas*, English Heritage, London (2006).
English Heritage, *Conservation Principles for the Sustainable Management of the Historic Environment: First Stage Consultation*, English Heritage, London (2006).
English Heritage, 'Heritage Protection Review', in *Conservation Bulletin*, English Heritage, London (Summer 2006).
English Historic Towns Forum, *Townscape in Trouble: Conservation Areas – The Case for Change*, English Historic Towns Forum, Bath (1992).
Esher, Viscount, *York: A Study in Conservation*, HMSO, London (1968).
Fawcett, Jane (ed.), *The Future of the Past*, Thames and Hudson, London (1976).
Feilden, Bernard, *Conservation of Historic Buildings*, Architectural Press, London (third edn, 2003); first published, Architectural Press, London (1982).
Fergusson, Adam and Mowl, Tim, *The Sack of Bath – And After*, Michael Russell, Salisbury (1989).
Fladmark, Magnus (ed.), *Cultural Tourism*, Donhead Publishing, Donhead St Mary (1994).
Fladmark, Magnus (ed.), *Heritage Conservation, Interpretation and Enterprise*, Donhead Publishing, Donhead St Mary (1993).
Geddes, Patrick, *Cities in Evolution*, with an introduction by Percy Johnson-Marshall, Ernest Benn, London (1968); first published by Williams and Norgate, London (1915).
Gehl, Jan and Gemzøe, Lars, *Public Spaces – Public Life: Copenhagen*, The Danish Architectural Press and The Royal Danish Academy of Fine Arts School of Architecture, Copenhagen (2004).

German Agency for Technical Cooperation (on behalf of the City Hall, Sibiu), *Charter for the Rehabilitation of the Historic Center of Sibiu/Hermannstadt*, GTZ, Sibiu (second edn, October 2000).
German Agency for Technical Cooperation (on behalf of the City Hall, Sibiu), *Kommunales Aktionsprogramm: Sibiu 2001–2004*, GTZ, Sibiu (2001).
German Agency for Technical Cooperation (on behalf of the City Hall, Sibiu), *Sibiu Historic Centre Management Plan: 2005–2009*, GTZ, Sibiu (2005).
Giovannoni, Gustavo, *L'urbanisme face aux villes anciennes*, with an introduction by Françoise Choay, Seuil, Paris (1998); first published as *Vecchie città ed edilizia nuova*, UTET Libreria, Rome (1931); second edn, CittàStudi Edizione, Rome (1995). Giovannoni first set out the principal elements of his thesis in a set of papers that were published in 1913 under the title *Vecchie città ed edilizia nuova: il quartiere del Rinascimento in Roma*.
Girardet, Herbert, *Creating Sustainable Cities*, Green Books, Totnes (1999).
Gravier, Jean-François, *Paris et le Désert Français*, Le Portulan, Paris (1947).
Hall, Peter, *The World Cities*, World University Library, Weidenfeld and Nicolson, London (1966).
Hall, Peter, *London 2000*, Faber and Faber, London (second edn, 1969); first published, Faber and Faber, London (1963).
Hall, Peter, *Cities of Tomorrow*, Blackwell, Oxford (updated edn, 1996); first published, Blackwell, Oxford (1988).
Harrison, Patrick (ed.), *Civilising the City: Quality or Chaos in Historic Towns*, Nic Allen, Edinburgh (1990).
Harrison, Patrick, *Urban Pride: Living and Working in a World Heritage City*, Edinburgh World Heritage Trust, Edinburgh (2002).
Haughton, Graham and Hunter, Colin, *Sustainable Cities*, Routledge, London (reprinted, 2003); first published, Jessica Kingsley (1994).
Hendry, Arnold, *The Edinburgh Transport Saga*, Whitehouse Publications, Edinburgh (2002).
Historic Environment Review Steering Group, *Power of Place: The Future of the Historic Environment*, Power of Place Office, English Heritage, London (2000).
Holliday, John (ed.), *City Centre Redevelopment*, Charles Knight, London (1973).
Howard, Ebenezer, *Garden Cities of To-morrow*, edited with a preface by F. J. Osborn and an introductory essay by Lewis Mumford, Faber and Faber, London (1946); first published as *To-Morrow: A Peaceful Path to Real Reform*, London (1898); second edn published as *Garden Cities of To-morrow*, Swan Sonnenschein, London (1902).
Insall, Donald and Associates, *Chester: A Study in Conservation*, HMSO, London (1968).
Insall, Donald, *The Care of Old Buildings Today: A Practical Guide*, Architectural Press, London (1972).
Jacobs, Jane, *The Death and Life of Great American Cities: The Failure of Town Planning*, Random House, New York (1961).
Jenks, Mike et al. (ed.), *The Compact City: A Sustainable Urban Form?*, E & FN Spon, London (1996).
Jodido, Philip, *Grands Travaux*, Connaissance des Arts, Paris (1992).
Johnson-Marshall, Percy, *Rebuilding Cities*, Edinburgh University Press, Edinburgh (1966).
Jokilehto, Jukka, *A History of Architectural Conservation*, Elsevier Butterworth-Heinemann, Oxford (2004); first published, Butterworth-Heinemann, Oxford (1999).
Kidder Smith, George, *The New Architecture of Europe*, World Publishing, New York (1961).
Kimmel, Alain (ed.), *Les Villes Nouvelles en Ile-de France*, Echos, Paris (1988).
Kostof, Spiro, *The City Shaped*, Thames and Hudson, London (1991).

Kostof, Spiro, *The City Assembled*, Thames and Hudson, London (1992).
Lacaze, Jean-Paul, *Paris: urbanisme d'État et destin d'une ville*, Flammarion, Paris (1994).
Latham, Derek, *Creative Re-Use of Buildings*, Donhead Publishing, Donhead St Mary (vols 1 and 2, 2000).
Le Corbusier, *The City of To-morrow and its Planning*, edited with a preface by Le Corbusier, Architectural Press, London (1947); first published as *Urbanisme*, Editions Crés, Paris (1924); first English edition published by John Rodker, London (1929).
Lowenthal, David, *The Past is a Foreign Country*, Cambridge University Press, Cambridge (1985).
Lowenthal, David, *The Heritage Crusade and the Spoils of History*, Cambridge University Press, Cambridge (1998).
Loyer, François, *Ville d'hier, ville aujourd'hui en Europe*, Editions du Patrimoine, Fayard, Paris (2001).
Loyer, François and Schmuckle-Mollard, Christiane, *Façadisme et Identité Urbaine*, Editions du Patrimoine, Paris (2001).
Lynch, Kevin, *The Image of the City*, MIT Press, Cambridge, Massachusetts (1960).
Lynch, Kevin, *Good City Form*, MIT Press, Cambridge, Massachusetts (1981).
McKean, John, *Giancarlo De Carlo: Layered Places*, Axel Menges, Stuttgart (2004).
Marchand, Bernard, *Paris, histoire d'une ville (XIXe –XXe siècle)*, Seuil, Paris (1993).
Matthew, Robert (ed.), *The Conservation of Georgian Edinburgh*, Edinburgh University Press, Edinburgh (1972).
Mehaffy, Michael (ed.), *The Order of Nature: The City as Ecosystem* (speaker position papers), The Prince's Foundation, London (September 2005).
Meller, Helen, *Patrick Geddes: Social Evolutionist and City Planner*, Routledge, London (1990).
Middleton, Michael, *Cities in Transition*, Michael Joseph, London (1991).
Ministry of Housing and Local Government, *The South East Study: 1961–1981*, HMSO, London (1964).
Mumford, Lewis, *The City in History*, Secker and Warburg, London (1961).
Newman, Peter and Kenworthy, Jeffrey, *Sustainablity and Cities: Overcoming Automobile Dependence*, Island Press, Washington DC (1999).
Niedermaier, Paul (ed.), *Nomination of the Sibiu Historic Centre for Inscription on the World Heritage List*, Institutul de Cercetări Socio-Umane Sibiu, Sibiu (September 2005).
Nordic World Heritage Office, *Sustainable Historic Cities: A North-Eastern European Approach*, Nordic World Heritage Office, Oslo (Final Report, December 1998).
Orbaşli, Aylin, *Tourists in Historic Towns: Urban Conservation and Heritage Management*, E & F N Spon, London (2000).
Oriolo, Leonardo (ed.), *Asmara Style*, Scuola Italiana, Asmara (1998).
Pearce, David, *Conservation Today*, Routledge, London (1989).
Petherick, Ann et al., *Living over the Shop: A Guide to the Provision of Housing above Shops in Town Centres*, Joseph Rowntree Foundation, York (first report, June 1990).
Pickard, Robert (ed.), *Management of Historic Centres*, Spon Press, London (2001).
Rasmussen, Steen Eiler, *London: The Unique City*, Jonathan Cape, London (revised edn, 1937); first published in Denmark (1934); abridged edn, Penguin Books, Harmondsworth (1960).
Rasmussen, Steen Eiler, *Towns and Buildings*, Liverpool University Press, Liverpool (1951); first published in Denmark (1949).
Richards, James (ed.), *European Heritage*, Phoebus Publishing, London (Issues One to Five, 1975).
Rodwell, Dennis, 'What does one do with a dilapidated pigeon coop?', in *Scottish Field*, Holmes McDougall, Glasgow (August 1986).

Rodwell, Dennis, 'The World Heritage Convention and the Exemplary Management of Complex Heritage Sites', in *Journal of Architectural Conservation*, Donhead Publishing, Donhead St Mary (November 2002).

Rodwell, Dennis, 'Industrial World Heritage sites in the United Kingdom', in *World Heritage Review*, UNESCO, Paris, and Ediciones San Marcos, Madrid (December 2002).

Rodwell, Dennis, 'Sustainability and the Holistic Approach to the Conservation of Historic Cities', in *Journal of Architectural Conservation*, Donhead Publishing, Donhead St Mary (March 2003).

Rodwell, Dennis, *The Bolshoi Theatre, Moscow, Russia* (mission report), Division of Cultural Heritage, UNESCO, Paris (2003).

Rodwell, Dennis, 'Approaches to Urban Conservation in Central and Eastern Europe' in *Journal of Architectural Conservation*, Donhead Publishing, Donhead St Mary (July 2003).

Rodwell, Dennis, *Over-arching Urban Planning and Building Conservation Guidelines for the Historic Perimeter of Asmara, Eritrea* (mission report), Cultural Assets Rehabilitation Project, Asmara (March 2004).

Rodwell, Dennis, 'Asmara: Conservation and Development in a Historic City', in *Journal of Architectural Conservation*, Donhead Publishing, Donhead St Mary (November 2004).

Rodwell, Dennis, 'Dubrovnik, Pearl of the Adriatic', in *World Heritage Review*, UNESCO, Paris, and Ediciones San Marcos, Madrid (December 2004).

Rodwell, Dennis, 'City of Bath: A Masterpiece of Town Planning', in *World Heritage Review*, UNESCO, Paris, and Ediciones San Marcos, Madrid (October 2005).

Rodwell, Dennis, 'Managing Historic Cities: the Management Plans for the Bath and Edinburgh World Heritage Sites', in *Journal of Architectural Conservation*, Donhead Publishing, Donhead St Mary (July 2006).

Rogers, Richard with Gumuchdjian, Philip, *Cities for a Small Planet*, Faber & Faber, London (1997).

Rogers, Lord (chairman), *Towards an Urban Renaissance*, Urban Task Force, London (1999).

Rogers, Richard and Powers, Anne, *Cities for a Small Country*, Faber & Faber, London (2000).

Rogers, Lord (chairman), *Towards a Strong Urban Renaissance*, Urban Task Force, London (2005).

Romanian-German Cooperation Project, *International Symposium on the Rehabilitation of Historic Cities in Eastern and South Eastern Europe* (symposium report), Sibiu City Hall and GTZ, Sibiu (2002).

Rosenau, Helen, *The Ideal City: Its Architectural Evolution in Europe*, Methuen, London (third edn, 1983); first published as: *The Ideal City in its Architectural Evolution*, Routledge and Kegan Paul, London (1959); second edn, Studio Vista, London (1974).

Samol, Frank, 'Urban Rehabilitation in Central and Eastern Europe', paper delivered at the *International Symposium on the Rehabilitation of Historic Cities in Eastern and South Eastern Europe*, Sibiu (October 2002).

Satterthwaite, David (ed.), *The Earthscan Reader in Sustainable Cities*, Earthscan Publications, London (1999).

SAVE Britain's Heritage, *Catalytic Conversion: Revive Historic Buildings to Regenerate Communities*, SAVE Britain's Heritage, London (1998).

Tetlow, John and Goss, Anthony, *Homes, Towns and Traffic*, Faber and Faber, London (second edn, 1968); first published, Faber and Faber, London (1965).

UNESCO, *Management of Private Property in the Historic City-Centres of the European Cities-in-Transition* (proceedings of UNESCO international seminar, Bucharest, April 2001), UNESCO, Paris (2002).

Ward, Barbara and Dubos, René, *Only One Earth: The Care and Maintenance of a Small Planet*, Norton, New York (1972).

Ward, Pamela (ed.), *Conservation and Development in Historic Towns and Cities*, Oriel Press, Newcastle-upon-Tyne (1968).

Watters, Diane and Glendinning, Miles, *Little Houses: The National Trust for Scotland's Improvement Scheme for Small Houses*, The Royal Commission on the Ancient and Historical Monuments of Scotland and The National Trust for Scotland, Edinburgh (2006).

Welter, Volker, *Biopolis: Patrick Geddes and the City of Life*, MIT Press, Cambridge, Massachusetts (2002).

Wilkinson, Philip, *Restoration: Discovering Britain's Hidden Architectural Treasures*, Headline, London (2003).

Wilkinson, Philip, *Restoration: The Story Continues*, English Heritage, Swindon (2004).

Williams, Katie et al. (ed.), *Achieving Sustainable Urban Form*, E & FN Spon, London (2000).

World Commission on Environment and Development, *Our Common Future* (known as the *Brundtland Report*), Oxford University Press, Oxford (1987).

Worskett, Roy, *The Character of Towns: An Approach to Conservation*, Architectural Press, London (1969).

Young, Michael and Willmott, Peter, *Family and Kinship in East London*, Penguin, London (revised edn with new introduction, 1986); first edn, Routledge and Kegan Paul, London (1957).

List of Figures

Chapter 1
1.1   Urbino, Italy: Ducal Palace 2
1.2   Fountains Abbey, England 3
1.3   St Albans Cathedral, England... 4
1.4   Bath, England: Queen Square 5
1.5   Brussels, Belgium: Grand-Place... 6
1.6   Edinburgh, Scotland: Sciennes Hill House, before restoration 9
1.7   Edinburgh, Scotland: Sciennes Hill House, after restoration............... 9
1.8   Dubrovnik, Croatia: the Stradun 11
1.9   Amsterdam, Netherlands 13
1.10 Sydney, Australia: Opera House 14
1.11 Marais quarter, Paris, France: Place des Vosges 15
1.12 Marais quarter, Paris, France: Hôtel le Rebours, street elevation 16
1.13 Marais quarter, Paris, France: Hôtel le Rebours, courtyard 16
1.14 Marais quarter, Paris, France: Hôtel de Sully, street elevation........... 17
1.15 Marais quarter, Paris, France: Hôtel de Sully, garden elevation........... 17
1.16 Avignon, France: *La Balance*, restored houses 18
1.17 Avignon, France: *La Balance*, new apartments 18
1.18 Plovdiv, Bulgaria: Ancient Reserve, Geordiady House 19
1.19 Plovdiv, Bulgaria: Ancient Reserve, derelict monument 20

Chapter 2
2.1   Dubrovnik, Croatia 24
2.2   Zamość, Poland 24
2.3   Zamość, Poland: residential courtyard 25
2.4   Rottingdean, England: Tudor-style cottages................................ 26
2.5   Darley Abbey, England: Brick Row.. 27
2.6   Paris, France: urban landscape 28
2.7   Ronchamp, France: Notre-Dame-du-Haut...................................... 29
2.8   Old Town, Edinburgh, Scotland: St Mary's Street 31
2.9   Old Town, Edinburgh, Scotland: Lawnmarket, courtyard elevation 32
2.10 Old Town, Edinburgh, Scotland: Lawnmarket, street elevation 32
2.11 Siena, Italy.. 34
2.12 Urbino, Italy 35
2.13 Bloomsbury, London, England: Woburn Square 36
2.14 Bloomsbury, London, England: Woburn Walk.................................. 36
2.15 Beverley, England: market square 38
2.16 Bath, England: Royal Crescent 40
2.17 York, England: medieval walls 42

List of Figures

2.18 York, England: riverside industries 43
2.19 Venice, Italy 45

Chapter 3
3.1 Kintzheim, France: La Montagne des Singes 48
3.2 Edinburgh, Scotland: Scott Monument 50
3.3 Ozone Layer 52
3.4 Eritrea, Horn of Africa 53
3.5 South Island, New Zealand 55
3.6 City of London, England: urban landscape 59
3.7 Paris, France: La Défense 61
3.8 Paris, France: La Défense, La Grande Arche 61
3.9 Paris, France: urban landscape with La Défense 62

Chapter 4
4.1 Belcastel, Dordogne, France 65
4.2 Florence, Italy 67
4.3 Ironbridge Gorge, England 68
4.4 Bath, England: Great Bath 68
4.5 Warsaw, Poland: market place 69
4.6 Rila Monastery, Bulgaria 70
4.7 Oslo, Norway: Gol stave church 71
4.8 Durham Cathedral, England 73
4.9 Etara, Bulgaria 74
4.10 Venice, Italy 76
4.11 Cambridge, England 76
4.12 Vilnius, Lithuania 78
4.13 Vilnius, Lithuania: tenement 79
4.14 Vilnius, Lithuania: tenement courtyard 79
4.15 Dubrovnik, Croatia: destroyed housing 80
4.16 Dubrovnik, Croatia: roof tiles 81
4.17 Dubrovnik, Croatia: old harbour 82
4.18 Dubrovnik, Croatia: housing 83
4.19 Dubrovnik, Croatia: children 83
4.20 Dubrovnik, Croatia: Gundulićeva Poljana, market 84

Chapter 5
5.1 Fatlips Castle, Scotland: exterior 87
5.2 Fatlips Castle, Scotland: interior 87
5.3 Shugborough Hall, England: entrance facade 89
5.4 Shugborough Hall, England: ornamental garden 89
5.5 Pittenweem, Scotland 90
5.6 South-East Scotland: the author's house 90
5.7 Glasgow, Scotland: Britannia Music Hall, street elevation ... 91
5.8 Glasgow, Scotland: Britannia Music Hall, auditorium interior 91
5.9 Darley Abbey, Derby, England: Darley Street 92
5.10 Darley Abbey, Derby, England: Brick Row, window 93
5.11 Darley Abbey, Derby, England: New Road, window 93
5.12 Darley Abbey, Derby, England: Poplar Row, door 93
5.13 Darley Abbey, Derby, England: Brick Row school, door 93
5.14 Darley Abbey, Derby, England: the Four Houses 94
5.15 Edinburgh, Scotland: Signal Tower, Leith 96
5.16 Kutna Horá, Czech Republic: craft businesses 97
5.17 Kutna Horá, Czech Republic: market square 99
5.18 Derby, England: disused upper floors 101

Chapter 5 (cont.)

5.19	Derby, England: former main post office	102
5.20	Saltaire, Bradford, England	104
5.21	Chartres, France: lower town and cathedral	106
5.22	Chartres, France: Rue des Ecuyers	107
5.23	Chartres, France: river Eure	109
5.24	Chartres, France: modern housing	109

Chapter 6

6.1	South-East England	113
6.2	Budapest, Hungary: Nyugati railway station	114
6.3	Sydney, Australia: Olympic Village, view from main stadium	115
6.4	Sydney, Australia: Olympic Village, railway station	116
6.5	Rothenburg-ob-der-Tauber, Germany	118
6.6	Rothenburg-ob-der-Tauber, Germany: Markusturn	118
6.7	Bratislava, Slovakia: statue of Napoleon	119
6.8	Vallingby, Stockholm, Sweden: town square	120
6.9	Paris, France: Georges-Pompidou Centre	123
6.10	Derby, England: former locomotive and carriage workshops	124
6.11	Copenhagen, Denmark: Nyhavn	125
6.12	Derby, England: market square	128
6.13	Marais quarter, Paris, France: Place des Vosges	129
6.14	Marais quarter, Paris, France: contemporary infill	130
6.15	Marais quarter, Paris, France: boutique	130
6.16	Ile-Saint-Louis, Paris, France: bakery	131
6.17	Ile-Saint-Louis, Paris, France: joiner and cabinetmaker	131

Chapter 7

7.1	Bath, England: Lansdown Crescent	137
7.2	Edinburgh, Scotland: view to the river Forth	137
7.3	Bath, England: Bath Abbey	138
7.4	New Town, Edinburgh, Scotland: Charlotte Square	139
7.5	Edinburgh, Scotland: Princes Street Gardens	140
7.6	Bath, England: Margaret Buildings	141
7.7	Bath, England: No 1 Royal Crescent	142
7.8	Bath, England: Beckford's Tower	143
7.9	New Town, Edinburgh, Scotland: Moray Place	145
7.10	New Town, Edinburgh, Scotland: Moray Place, window balcony	147
7.11	New Town, Edinburgh, Scotland: Heriot Row, street lantern	147
7.12	New Town, Edinburgh, Scotland: Ann Street	148
7.13	Dean Village, Edinburgh, Scotland	150
7.14	Old Town, Edinburgh, Scotland: Greyfriars Bobby	151
7.15	New Town, Edinburgh, Scotland: No 17 Heriot Row	151
7.16	Bath, England: Green Street	152
7.17	Bath, England: Bath Spa Building	155
7.18	Edinburgh, Scotland: the Scottish Parliament	156

Chapter 8

8.1	Budapest, Hungary	162
8.2	Sibiel, Transylvania, Romania	163
8.3	Sibiu, Romania: inner fortifications	164
8.4	Sibiu, Romania: Piața Mare	165
8.5	Sibiu, Romania: upper to lower towns	166
8.6	Sibiu, Romania: Avram Iancu street	167
8.7	Sibiu, Romania: Avram Iancu street, courtyard housing	167
8.8	Sibiu, Romania: Piața Mică, restaurant	168

8.9   Banffy Castle, Bonțida, Romania 169
8.10 Banffy Castle, Bonțida, Romania 169
8.11 Eritrea, Horn of Africa 171
8.12 Asmara, Eritrea: Fiat Tagliero service station 173
8.13 Asmara, Eritrea: Harnet Avenue 174
8.14 Asmara, Eritrea: market square 174
8.15 Asmara, Eritrea: Medeber industrial area 175
8.16 Asmara, Eritrea: urban landscape 177
8.17 Asmara, Eritrea: joinery workshop 178
8.18 Asmara, Eritrea: villa 179
8.19 Asmara, Eritrea: courtyard housing 179
8.20 Asmara, Eritrea: Selam Hotel 181

Chapter 9
9.1   San Gimignano, Italy 184
9.2   Montepulciano, Italy: window 186
9.3   Urbino, Italy: window 186
9.4   Brașov, Romania: urban landscape 187
9.5   Athens, Greece: the Erechtheum 188
9.6   Cheltenham, England: Montpellier Walk 188
9.7   Dryburgh Abbey, Scotland 190
9.8   Melrose Station, Scotland: derelict interior 191
9.9   Melrose Station, Scotland: new restaurant 191
9.10 Melrose Station, Scotland: restored exterior 191
9.11 Bolshoi Theatre, Moscow, Russia: exterior 193
9.12 Bolshoi Theatre, Moscow, Russia: auditorium interior 193
9.13 Darley Abbey Mills, Derby, England 194
9.14 Darley Abbey Mills, Derby, England: cotton preparation range 195
9.15 Zamość, Poland: market square 197
9.16 Paris, France: *bouquinistes* 197
9.17 Banská Štiavnica, Slovakia 198
9.18 Prague, Czech Republic 199
9.19 Sighișoara, Transylvania, Romania 200
9.20 St Petersburg, Russia: Catherine Palace, Tsarskoie Selo, exterior 201
9.21 St Petersburg, Russia: Catherine Palace, Tsarskoie Selo, Great Hall . 201
9.22 St Petersburg, Russia: General Staff Headquarters 202
9.23 St Petersburg, Russia: housing 202

Chapter 10
10.1 Glasgow, Scotland: facadism 207
10.2 Moscow, Russia: high-rise block 209
10.3 Helsinki, Finland: company headquarters 209
10.4 Nuremberg, Germany: urban landscape 210
10.5 Nuremberg, Germany: restored and new buildings 210
10.6 Nuremberg, Germany: new buildings 210
10.7 Kutna Horá, Czech Republic: former Jesuit College 211
10.8 Bath, England: Pulteney Bridge 212
10.9 Culzean Castle, Scotland 212
10.10 Pécs, Hungary: contemporary intervention 213
10.11 Pécs, Hungary: public art 213
10.12 Paris, France: Louvre Pyramid 214

Index

Numbers in *italics* refer to Figures

Aalto, Alvar, 209
Abercrombie, Sir Patrick, 140, 141, 153, 154
acid rain, 47, 49
Adam, Robert, 139, 212–13
Africa, 54, 171, 172, 173
Agenda 21, 54, 55, 196
amenity societies/groups, 88, 110, 141, 142–43, 144
America, North, 26, 51
Amsterdam (Netherlands), *1.9*
Ancient Monuments Act (1882), 86, 87
Annan, Kofi, 52
anthropological approach/vision, 74, 84, 187, 206
anti-urban movement/legacy/tradition, 25–26, 45, 216
ARCH (Art Restoration for Cultural Heritage) Foundation, 83
architect: word derivation, 211
architects, integrated training, 35, 211
architectural and historic interest, 1, 6, 21, 58, 63, 91, 95, 98, 116, 188, 198, 227
architectural conservation, vii, ix, 20, 56, 57, 63, 72, 146–48, 158, 166, 175, 180, 195
    beginnings and evolution, 1–6, 21, 23, 67, 185–87
    charters, 10–14
    language, 7–9, 21–22, 58
    linkage to sustainability, 204, 205–206, 207, 211, 215
    potential, 187
architectural integrity, 8, 9
Asia, 51, 70
Askins, Dave, 168
Asmara (Eritrea), 170–82, 196; *8.12, 8.13, 8.14, 8.15, 8.16, 8.17, 8.18, 8.19, 8.20*
Athens (Greece), *9.5*
Athens Charter (1931), 10, 12, 36
Attenborough, Sir David, 48–49

authenticity:
    concept, 1, 4, 6, 8, 10, 12, 21–22, 85, 186, 188
    ICCROM definition, 8–9, 71–72, 85
    inclusive approach, 69–73
    loss (United Kingdom), 110
    World Heritage criteria, 94, 134, 136, 159
Avignon (France), La Balance, 18; *1.16, 1.17*

Baltic States, 77
Bandarin, Francesco, 214
Banffy Castle (Bonțida, Romania), 169–70; *8.9, 8.10*
banlieue: word derivation, 18
Banská Štiavnica (Slovakia), *9.17*
Barcelona (Spain), 122
Bath (England):
    Abbey, *7.3*
    Bath Spa Building, *7.17*
    Beckford's Tower, *7.8*
    Green Street, *7.16*
    Lansdown Crescent, *7.1*
    Margaret Buildings, *7.6*
    Pulteney Bridge, *10.8*
    Queen Square, 39, 139; *1.4*
    Roman Baths, *4.4*
    Royal Crescent, *2.16, 7.7*
    Save Bath campaign, 5, 18, 41
    Study in Conservation (1968), 38–41, 44, 46
    World Heritage Site, 69, 75, 133, 136–60
Bath City Council, 144
Bath and North East Somerset Council, 146, 158
Bath Preservation Trust, 142–43
Bergen (Norway), Bryggen quarter, 71
Beverley (England), *2.15*
Bhopal (India), 49
biodiversity, 47, 48, 54, 55, 63, 74, 113, 183, 185
biosphere, ix, 52, 57, 183

Blaenavon Industrial Landscape (Wales), 74–75
Bolshoi Theatre (Moscow, Russia), 192–93; *9.11*, *9.12*
bottom-up, 46, 63, 135, 161–82, 195–97, 203, 207
Braşov, Romania, 170; *9.4*
Bratislava (Slovakia), *6.7*
Britain: see United Kingdom
British Council in Romania, 170
brownfield land/site, 117, 121, 123–24
Brundtland, Gro Harlem, 52
Brundtland Report (1987) (Our Common Future), 52–53, 56, 63, 111, 157, 183
   sustainable development, definition, 56
Brussels (Belgium), Grand-Place, *1.5*
Buchanan, Sir Colin (and Partners), 5, 39–41, 44, 141, 152, 154
   see also Bath, Study in Conservation
Buchanan Report, 37–38, 46, 114
Bucharest (Romania), 162
Budapest (Hungary), 161; *6.2*, *8.1*
building preservation trust/s (UK), 88, 144
Bulgarian Renaissance, 19, 70
Burra Charter (1999 revision), 8, 14, 21, 77, 104, 157, 180
   definitions, 8

Cadw (historic environment division, Wales), 86
Cambridge (England), 75; *4.11*
Carcassonne (France), 72
CARP (Cultural Assets Rehabilitation Project, Eritrea), 171, 173, 175, 176
Český Krumlov (Czech Republic), 99
character and appearance, 8–9, 91, 95, 227
character appraisals, 100
charitable trusts, 88, 110
Charte d'Athènes (1933), 10, 12, 28
Chartres (France), 106–109, 207; *5.21*, *5.22*, *5.23*, *5.24*
Cheltenham (England), *9.6*
Chernobyl (Ukraine), 49
Chester (England), 75
   Study in Conservation (1968) (Insall report), 38, 41, 44, 46
Chichester (England)
   Study in Conservation (1968), 38, 41–42, 46
China, 50, 75, 203
Civic Amenities Act (1967), 39, 40, 86, 95, 227
Civic Trust (England and Wales), 88, 90, 194
classical antiquity, 1–2, 3, 213
Clean Air Acts, 50
climate change, 47, 54, 63, 183
Club of Rome, 51
Cockburn Association (Edinburgh, Scotland), 144
Cockburn, Lord, 144

Cologne (Germany), 208
community:
   belonging, 103, 110, 117, 122, 185, 187, 205
   involvement, 98, 168–70, 181
congestion charging, 125, 153
Congress of Amsterdam, 13
conservation:
   appraisals, 105
   architects' role, 211
   Burra definition, 8
   charters, 6, 7, 10–14, 21, 208, 215
   coincidence with sustainability, 183–203, 207, 216
   concept, vii, ix, 7–8, 22, 47, 55, 56, 104, 129, 163, 215
   deficit, 110
   ethic, 88, 98, 158
   international initiatives, 64–85
   Jokilehto, 205–206
   living: concept, 33
   officers, 95, 100
   plans, 14
   surgery, 31–32
conservation areas (United Kingdom), 39, 86, 87, 91, 94, 95, 96, 98, 103, 105, 106, 110, 227
contemporary:
   architecture/building, 10, 11, 12, 110, 129–30, 155–56, 208–11, 214
   construct, 208, 213
   Rila Monastery (Bulgaria), 70
Copenhagen (Denmark), 125–26, 154; *6.11*
Council of Europe, 5, 12, 133, 163
craft skills, 6, 21, 23, 96, 100, 102, 106, 110, 148, 166, 169, 186, 187, 195, 206, 215
creative continuity, 1, 10, 74, 84, 212–13
creative skills, shortfall, 211
Cullen, Gordon, 20–21, 44
   urban design, definition, 21, 22, 209, 211
Culross (Scotland), 90
cultural:
   continuity, 1, 175, 187, 206
   diversity, ix, 2, 21, 30, 51, 59, 63, 71, 74, 84, 98, 113, 134, 176, 185–86
   identity, vii, 2, 30, 46, 54, 59, 69, 92, 134, 185–86, 187, 206, 208, 209, 210
   significance, 8, 14, 100, 104, 110, 166, 188, 189, 215
cultural landscape, 67–69, 84, 136, 206
culture: evolving concept, 73, 188
Culzean Castle (Scotland), *10.9*
Czech Republic, 97, 99

Darley Abbey (Derby, England):
   mills, 194–95; *9.13*, *9.14*
   village, 27, 92–94; *2.5*, *5.9*, *5.10*, *5.11*, *5.12*, *5.13*, *5.14*

Darwin, Charles, 30
De Carlo, Giancarlo, 35
Declaration of Amsterdam (1975), 12–13

English Heritage, 7, 86, 87, 94, 95,
  policy review, 98–106, 110
  regeneration agency, 104–105
English Historic Towns Forum, 95
Enlightenment, Age of, 1, 31, 156
environmental:
  area, 37, 39, 184
  assessments, 55
  awareness, 51
  capacity, 37–38, 41, 135, 152, 154, 184
  capital, 100, 114, 126, 132, 185, 189, 203, 205
  impact, 185
  performance, 10, 166, 185, 189, 203, 215
  protection, 53
  responsibility, 51, 55, 126, 132, 185
  stakeholder, 185, 196, 203
equity, global/social/inter-generational, 47, 53, 55, 56, 113, 183, 184, 205
Eritrea, 170–71, 182; *3.4*, *8.11*
Esfahan (Iran), 208
Esher, Lord: *see* York, Study in Conservation
Etara (Bulgaria), *4.9*
Europa Nostra, 147
Europe, vii, 5, 10, 15, 27, 29, 55, 60, 70, 72, 75, 97, 112
  Central and Eastern, 10, 77, 96, 161, 198, 199–201, 207, 208
  Continental, 26, 117, 127–28
  East, 23, 127
  Western, 161, 163, 164, 180
European Architectural Heritage Year (1975), 5–6, 12–13, 88, 205
European Charter (1975), 10, 12–13, 21, 99, 133
European Union agricultural policy, 112–13
evolution/ary, 30, 32, 35, 36, 46, 57, 180, 182, 183, 196
Eyck, Aldo van, 44

facadism, 135, 207
Fatlips Castle (Scotland), *5.1*, *5.2*
Feilden, Sir Bernard, 8, 57
Florence (Italy), *4.2*
Foster, Norman, 214
Fountains Abbey (England), *1.2*
France, 15, 18, 22, 26, 30, 60, 62, 106, 162, 208
French Revolution, 4
Friends of the Earth, 51

Garden City, 27–29, 117, 120, 172
Geddes, Sir Patrick, 29–33, 36, 46, 47, 57, 58, 111, 186
Gehl, Jan, 125, 154
gentrification, 165
geocultural diversity: *see* cultural diversity

Germany, 162
Giovannoni, Gustavo, 30, 33–36, 46, 58, 63, 140, 186, 189, 208, 211
Glasgow (Scotland):
  Britannia Music Hall, *5.7*, *5.8*
  Building Preservation Trust, 91
  facadism, *10.1*
global warming, 47, 49, 54
Graz (Austria), Kunsthaus, 214
Green Belt, 142, 144
  Act, 27
greenfield land/site, 112, 115, 117, 126–27, 189
Greenpeace, 51, 56, 115
GTZ (Deutsche Gesellschaft für Technische Zusammenarbeit) (German Agency for Technical Cooperation), 163–68

Habsburg, Archduchess Francesca von, 83
Hall, Sir Peter, 59, 127
harmonious coexistence, 34, 208
Harrap, Abigail, 146
Haussman, Baron, 131
Helsinki (Finland), *10.3*
Hendry, Arnold, 154
heritage:
  assets: definition, 103, 105
  champions, 102
  construct, 7, 74, 85, 89, 100, 206, 208, 215
  definitions, 7, 22, 66, 67, 84, 116
  industry, 7
  urban: concept, 33
Herschel, William, 143
historic environment:
  catalyst, 182
  concept/definition, 98–106, 206, 228
  inclusive view, 188–89
historic landscape, 98, 100
Historic Scotland, 7, 86, 146
Holland: *see* Netherlands
home zones, 124, 180
housing, 14, 41, 43, 83, 99, 125, 161–62, 163–68, 177–79, 192
Howard, Sir Ebenezer, 27–29, 30, 45, 117, 120
Hungary, 161
Huxley, Aldous, 47
Huxley, Sir Julian, 47
Huxley, Thomas, 30, 47

ICCROM (International Centre for Conservation in Rome; officially, International Centre for the Study of the Preservation and Restoration of Cultural Property), 8, 66–67, 77, 159
ICCROM Management Guidelines (1998), 135–36, 149

ICOMOS (International Council on Monuments and Sites), 12, 14, 66, 72, 134, 159
  advisory report, 70, 72, 73, 145
iconic buildings/architecture, 100, 206
industrial accidents, 47, 49
Industrial Revolution, 25, 30, 45, 47, 55, 74, 114
Insall, Donald, 196; *see* also Chester
Institute of Historic Building Conservation, 169
intangible cultural heritage: UNESCO agenda, 73–75, 136, 186–87
integrity:
  concept, 73, 85, 186
  loss (United Kingdom), 110
  UNESCO definition, 73
  World Heritage criteria, 94, 134, 136, 159
International Commission on Sustainable Development, 54
interpretation: concept, 66, 85
Ironbridge Gorge (England), 69; *4.3*
Italian Renaissance, 1–2, 3, 10, 23–24, 27, 67, 111, 213
Italy, 30, 35, 46, 75, 106, 170, 172, 189
IUCN (International Union for Conservation of Nature and Natural Resources), 48, 67
  Red List of Threatened Species, 48, 67

Jacobs, Jane, 196
Japan, 71
Jenkins, Dame Jennifer, 106
Johannis, Mayor Klaus, 168, 170
John Paul II, Pope, 53
Jokilehto, Jukka, 205, 215

Kinzheim (France), Montagne des Singes, *3.1*
Krakov (Poland), Kazimierz quarter, 148
Krier, Leon, 117
Kutna Horá (Czech Republic), *5.16*, *5.17*, *10.7*
Kyoto Protocol (1997), 54, 63

Landmark Trust, 211
Le Corbusier (Charles-Edouard Jeanneret), 28–29, 30, 34, 45, 46, 60, 114
Letchworth Garden City (England), 27
Little Houses Improvement Scheme (Scotland), 88, 90
Liverpool (England), 75, 103, 124
Local Authorities World Heritage Forum, 101
local distinctiveness, 91–94, 95, 97, 98, 103, 110, 185, 196, 215
London (England), 36, 117, 119, 127, 189
  Bloomsbury, *2.13*, *2.14*
  City, 36, 59, 208; *3.6*
  ecological footprint, 112
  metropolitan city, 59–60

smogs, 49–51
urban landscape (skyline), 59–60
LOTS (Living Over The Shop project), 127
Lynch, Kevin, 21

Malraux, André, 30
Malraux, Loi (1962), 15, 30, 107, 108
Malthus, Thomas, 47
managing change/management of change, 44, 100, 103, 104, 110, 149, 155
Manchester (England), 124
Martin, Sir Leslie, viii
Melrose Station (Scotland), 191; *9.8*, *9.9*, *9.10*
metropolitan city/ies, 29, 59–63, 128–31, 189, 204
Miller, John, 144, 191
minimum intervention, 12, 104, 110, 135, 168, 180, 182, 186, 189, 194–95, 207, 211, 215
misspent wealth, x, 182, 198, 203
mixed-use, 113, 117, 119, 120, 122, 124, 126, 132, 161
Modern Movement, 3, 5, 10, 12, 21, 27–28, 29, 208
monocentric city, 59–60, 121, 208
Montepulciano (Italy), *9.2*
Montreal Protocol (1987), 52, 54, 63, 75
monument:
  concept, 1, 3, 12, 21
  definition, 10
Morris, William, 4, 26
Moscow (Russia), *10.2*
museological approach, 15–20, 22, 35, 129

Napoleon I, 119, 131
Napoleon III, 131
Nara Document on Authenticity (1994), 71–72, 85
National Trust (England, Wales, Northern Ireland), 88–89
National Trust for Scotland, 88, 90, 212
Netherlands, 106
New Deal for Communities (England), 122
New Lanark (Scotland), 26
New Towns (England), 60
  Act, 27
New Urbanism movement (United States of America), 117
New Zealand:
  South Island, *3.5*
  Tongariro National Park, North Island, 67–68
Nîmes (France), Carré d'Art, 214
non-conforming uses, 43–44
Norwich (England), 37–38, 75
Nuremberg (Germany), 210; *10.4*, *10.5*, *10.6*

OPEC (Organisation of Petroleum Exporting Countries), 51
Orwell, George, 47

Oslo (Norway), Gol stave church, *4.7*
OTRA (Old Town Renewal Agency): *see* Vilnius
outstanding universal value, 65, 77, 150
Oxford (England), 75
ozone layer/depletion, 47, 49, 52, 63, 183; *3.3*

Palladian, 5
Palladio, Andrea, 139
Paris (France), 117, 118, 131
  bouquinistes, *9.16*
  Georges-Pompidou Centre, 214; *6.9*
  Ile-Saint-Louis, *6.16*, *6.17*
  La Défense, 60–63, 208; *3.7*, *3.8*
  Louvre Pyramid, 214; *10.12*
  Marais quarter, 15–18, 28, 60, 123, 128–31, 208; *1.12*, *1.13*, *1.14*, *1.15*, *6.14*, *6.15*
  metropolitan city, 60–63, 128–31, 189, 207–208
  Place des Vosges, 208; *1.11*, *6.13*
  Plan Voisin (1925), 28, 60
  urban landscape/skyline, 60, 62–63; *2.6*, *3.9*
pastiche, 18, 135, 155, 211, 213
Pathfinder (Housing Market Renewal Initiative, England), 126, 127
Pécs (Hungary), *10.10*, *10.11*
pedestrianisation, 6, 41, 43, 107, 125, 152, 153, 178
Pei, Ieoh Ming, 214
permitted development rights, 95, 105, 159
photochemical smogs, 51
picturesque: definition, 3
Picturesque Movement, 1, 3, 21, 87
Pittenweem (Scotland), *5.5*
Playfair, William, 140
Plovdiv (Bulgaria), Ancient Reserve, 19–20; *1.18*, *1.19*
Plymouth (England), 5
Poland, 162
pollution, air/noise/urban, 25, 45, 47, 49–51, 59, 63, 112, 113, 132, 152, 178, 183, 184, 203
polycentric city, 62–63, 117, 121, 123, 208
Poundbury (Dorchester, England), 117
Prague (Czech Republic), 199; *9.18*
pre-industrial city/ies, 23–25, 26, 29, 33, 46, 114, 132, 205
Prescott, John, 214
presentation, concept, 65, 66
preservation:
  Burra definition, 8
  concept, 7–8, 22
preserve and enhance/preservation and enhancement, 39, 91, 95, 227
preventative maintenance, 100, 158
protection: concept, 12, 64, 65–66, 85
proximity, 23, 31, 34, 38, 39–40, 44, 113, 118, 121, 122, 131, 132, 195, 208

public art, 151, 180, 213
PVC (polyvinyl chloride), 115, 116, 168, 170

quality of life, ix, 49, 57, 58, 103, 183

Rauma (Finland), Old, 71
reconstruction: Burra definition, 8
recycle/recycling/recycled, 115–16, 121, 123, 147, 174–75, 187–92, 203, 204, 215
reduce, 116, 187–92, 203
regeneration, 78–79, 101, 103, 104–105, 108, 121–24, 127, 129, 148, 194
resource/s:
    material, 58, 59, 110, 112, 132, 168, 181, 183, 185, 189, 204, 206–207, 216
    natural, ix, 47, 51, 57, 58, 63, 112, 114, 185, 204
    non-renewable, 56, 63, 103, 112, 114, 116, 132, 183, 187, 195, 204, 207, 215
restoration:
    Burra definition, 8
    concept, 7–8, 22
    stylistic, 4, 12, 72–73, 85
    Viollet-le-Duc definition, 4
reuse/reusable, 41, 116, 187–95, 203, 204, 215
Rila Monastery (Bulgaria), 70, 72; *4.6*
Rio Earth Summit/Rio Declaration (1992): *see* UN
road pricing: *see* congestion charging
Rodwell, Dennis (architect), 8–9, 31, 32, 90, 91, 96, 190, 191
Rogers, Richard (Lord), 121, 122–23, 124, 127
Romania, 112, 162, 163, 169–70
Romanian-German Cooperation project: *see* GTZ
Romantic Movement, 2, 21
Rome, 190
Ronchamp (France), *2.7*
Rothenburg-ob-der-Tauber (Germany), *6.5*, *6.6*
Rotterdam (Netherlands), 5
Rottingdean (England), *2.4*
Rowling, J K, 157
Ruskin, John, 26

Sabaudia (Italy), 172
St Albans Cathedral (England), *1.3*
St Petersburg (Russia), 199–203, 208; *9.22*, *9.23*
    Marinsky Theatre, 214
    Tsarskoie Selo, Catherine Palace, *9.20*, *9.21*
Saltaire (Bradford, England), 26; *5.20*
San Gimignano (Italy), *9.1*
Save Britain's Heritage, 126
Schusev, Alexander, 213
Scotland, 30, 46, 88
    cities, 127–28
    South-East/Scottish Borders, 190; *5.6*

Scott, Sir George Gilbert, 4, 8
Scott, Sir Peter, 48
Scott, Sir Walter, 50, 192
Scottish Civic Trust, 88
Second World War, 5, 10, 15, 29, 30, 35, 40, 47, 63, 70, 119, 140–41, 143, 161, 169–70, 199, 210
secteurs sauvegardés (France), 15–18, 22, 107
sense of:
    belonging: *see* community belonging
    place, 20, 23, 110, 119, 136, 205
set-aside policy, 112–13
Shanghai (China), 203
Shankland, Graeme, x, 198
Shelley, Mary, 47, 222
Shugborough Hall (England), *5.3*, *5.4*
Sibiel (Romania), *8.2*
Sibiu (Romania), 162–68, 170, 182, 196; *8.3*, *8.4*, *8.5*, *8.6*, *8.7*, *8.8*
    Rehabilitation Charter, 164–65
Siena (Italy), *2.11*
Sighişoara (Romania), 199–200; *9.19*
Sites Law (France, 1930), 60, 65
skyline: *see* urban landscape
slum clearance, 28, 29, 33, 41, 92, 104, 114, 126, 141, 196, 220
Smith, Sir John, 211, 215
social cohesion/inclusion, 98, 110, 185, 205
Society of Antiquaries, 73
Sorlin, François, 16
SPAB (Society for the Protection of Ancient Buildings), 4, 72, 88, 144
SPAB Manifesto (1877), 4, 10, 12, 26, 87
statements of:
    importance, 105
    significance, 14
Stevens, Sir Jocelyn, 105
Stockholm (Sweden), 120
    Vallingby, *6.8*
strategic:
    approach, 63, 128, 144, 189, 207–208
    plan, 176
sustainability:
    agenda, 76, 98, 103, 110, 132, 136, 159, 215
    beginnings, 47–56
    coincidence with conservation, 183–203, 204, 207, 216
    concept, vii, ix, 63, 64, 74, 103, 136, 149, 160, 183–85
    Jokilehto, 205–206
    language, 56–58
    relevance to historic cities, 58–59
    revolution, 56

sustainable:
  evolution: concept, 57
  growth, 53
  management, 189
  regeneration, 99
  relationship, 114
  over-usage, 56–57
  world, 112, 132, 204
sustainable city/ies:
  concept and vision, 111–16, 120–21, 122, 123, 131–32, 189, 205, 208
  definition, 111, 186
sustainable community/ies (United Kingdom), 57–58, 105, 123, 124
sustainable development:
  architects' role, 211
  Brundtland definition: 56–57, 183
  concept, vii, 52–54, 64, 74, 103, 111, 119, 122, 124, 132, 135, 148, 159, 163, 171, 175, 179, 181–82, 205, 208, 216
Sydney Australia,
  Olympic Village (2000), 115–16; 6.3, 6.4
  Opera House, 1.10

Telč (Czech Republic), 99
Telford Schools World Heritage Project (England), 168
theatrical scenery: see urban stage set
Third World, 53
top-down, 46, 63, 135, 181, 195–97, 203
tourism, 7, 39, 99, 103, 149, 171
tourist industry, 81–82, 98, 100
town planning: see urban planning
townscape:
  concept, ix, 20–22, 23, 41, 44–45, 58, 63, 95, 154, 157
  Cullen definition, 20
townscape appraisals, 44, 100
Town and Country Planning Act (1944), 86
Transylvania, 162, 163
Transylvania Trust (Romania), 169

United Kingdom, vii, 1, 5, 8, 25–26, 27, 29, 46, 58, 65, 75–76, 112, 113, 117, 121, 129, 189, 208, 215, 216
  conservation position and directions, 86–110
  construction industry, 102, 188, 206
  ecological footprint, 112
  interpretation, 66, 85
  managing World Heritage Cities, 133–60
  urban conservation beginnings, 36–44
  urban pollution, 49–50

UN (United Nations):
  Conference on Environment and Development (1992) (Earth Summit, Rio de Janeiro, Brazil), 54, 63, 183, 196
  Conference on the Human Environment (1972) (Stockholm, Sweden), 52, 54, 64
  Environment Programme, 52
  sustainability initiatives, 52–55, 64, 74, 75
  World Summit on Sustainable Development (2002) (Earth Summit, Johannesburg, South Africa), 54
UNESCO (United Nations Educational, Scientific and Cultural Organisation), 56, 64–85, 159, 163, 186, 208; see also World Heritage
  Bolshoi project, 192–93
  definition of heritage, 7, 84, 116
  Intangible Cultural Heritage Convention (2003), 74, 84
  Operational Guidelines (2005), 64, 69, 73, 94, 134, 136, 158, 159
  Recommendation (1972) (national), 66, 75, 84, 133
  Recommendation (1976) (historic areas), 134
United States of America: National Park Service, 65
upper floors, vacant/disused/reuse (England), 39, 40–41, 42, 101, 102, 127–28, 132, 189
urban conservation:
  charters, 10–14, 21
  concept, vii, ix, 56, 132, 135, 142, 146, 181, 185–87
  expanding threats, 198–203
  integrated, 13, 77–78, 98–99, 101, 105, 110, 133
  international cooperation, 77–85
  linkage to sustainability, 204, 205–206, 207, 215, 216
  museological beginnings, 15–20, 22
  strategic beginnings, 59–63
  United Kingdom beginnings, 36–44
urban design, 20, 21, 22, 41, 100, 123, 132, 145, 209
  Cullen definition, 21
urban landscape, 20, 28, 59–60, 62–63, 65, 140, 162, 176–77, 179, 187, 200, 202, 208, 210
urban planning, 13, 15, 20, 21, 22, 47, 58, 66, 122, 134, 139, 145, 146, 154, 195–96
  Asmara guidelines, 170, 172, 178, 179, 180, 181–82
  background, 23–36, 45–46
urban renaissance, 121–27, 132, 214, 215
urban stage set, 40, 44, 196
Urban Task Force, 99, 101, 102, 121–24, 127, 128
urban village, 117–19, 126, 132, 176
Urban Villages Group, 117, 120

Urban Villages report, 117, 119
Urban White Paper, 128
Urbino (Italy), *1.1*, *2.12*, *9.3*

VAT (Value Added Tax), 97
Venice (Italy), *2.19*, *4.10*
Venice Charter (1964), 10–11, 12, 13, 72
Vienna (Austria), 208
Vilnius (Lithuania), 77–79, 85, 148; *4.12*, *4.13*, *4.14*
Viollet-le-Duc, Eugène Emmanuel, 4, 8, 18, 72, 157

Wales, Prince of, 117, 156
Warsaw (Poland), 70; *4.5*
Washington Charter (1987), 11, 13–14, 21, 134, 155
Waverley Route (Scotland), 153, 192
Welwyn Garden City (England), 27
Wood, John:
  elder, 5, 140
  younger, 40, 142
World Bank, 77
World Commission on Environment and Development: *see* Brundtland Report
World Conservation Union: *see* IUCN
World Heritage:
  brand, 75–77
  Centre, 66, 133, 214
  Committee, 66, 70, 71, 74, 75
  List, 65, 66–67, 74, 75, 77, 85, 133, 136, 144, 159
  List in Danger, 65, 77, 81
  Tentative List, 150, 163, 172
World Heritage Convention (1972), 2, 64–66, 67, 71, 74, 75, 77, 79, 84, 94, 101, 133, 163, 172
  global strategy, 67, 75–76
  state parties, 67, 85, 133
World Heritage Sites, 103, 133, 136
  buffer zone, 134, 144, 145
  management plans, 105, 133–60
  monitoring, 158–59
  periodic reports, 133, 158
  state of conservation, 133, 158
World Monuments Fund Watch List, 199
world population, 47, 51, 63, 112
Worskett, Roy, 44
WWF (World Wildlife Fund), 48
Wyatt, James, 73

York (England), 75; *2.17*, *2.18*
  Study in Conservation (1968) (Esher report), 38, 42–44, 46, 75, 76, 96

Zamość (Poland), *2.2*, *2.3*, *9.15*
zone/zoning, 19, 27, 46, 127, 177–78